The Book of Five Keys

A Cyber Security Interpretation of Miyamoto
Musashi's Book of Five Rings

The Book of Five Keys draws foundational inspiration from Miyamoto Musashi's The Book of Five Rings, a classic treatise on strategy and combat. This original work, first published in the 17th century, is in the public domain in its country of origin and in other jurisdictions where copyright has expired; specifically, those recognizing copyright as the author's life plus 100 years or fewer. It is also in the public domain in the United States, as it was published prior to January 1, 1930. This open-use status enabled this author to freely reinterpret and recontextualize Musashi's timeless principles into a modern framework tailored for cybersecurity, creating a fusion of ancient strategic wisdom and contemporary digital defense.

For more information or to book an event, contact:
Bob@c-ooda.com
http://www.c-ooda.com

ISBN: 979-8-9881956-3-4
Softcover First Edition: May 2025

Editor – Marya Roddis, CTPRP, S.U.N. Resource Development

Contents

About the Author

Bob Maley is a seasoned Chief Information Security Officer and strategic advisor known for transforming how organizations think about cyber risk. With over two decades of experience at the forefront of cybersecurity, he has led global third-party risk programs, pioneered AI-driven risk models, and guided companies through some of the world's most complex regulatory environments.

Bob's approach to cybersecurity is rooted in clarity, precision, and adaptability – principles that echo Miyamoto Musashi's timeless strategy. Rather than relying on stale frameworks or checkbox compliance, Bob champions fluid thinking, outcome-driven defense, and continuous adaptation through models like FAIR and the OODA loop.

He has spoken on stages worldwide, led programs conducting over 1,400 annual vendor assessments, and helped executive teams translate cyber chaos into business-aligned decisions. But beyond the certifications and C-suite titles, Bob is a strategist at heart — someone who sees cybersecurity not as a technical field, but as a modern battlefield requiring discipline, timing, and intent.

This book represents a fusion of his philosophy: that cybersecurity—like swordsmanship—is less about the weapon and more about the warrior's mindset.

Prologue

"In order to win, you should operate at a faster tempo or rhythm than your adversary."

— Col. John Boyd

The Genesis of the Book of Five Keys: From Ancient Battles to Digital Warfare

"Strategy without tactics is the slowest route to victory. Tactics without strategy is the noise before defeat." – Sun Tzu

Have you ever played a video game that changed your life? Yeah, me neither. Ok, maybe a few.

But in 1984, I discovered a computer game from Broderbund software called "The Ancient Art of War." For a technology geek like me, it was like finding Excalibur in a stone...or, well, a floppy disk. This game wasn't just about bashing pixels together; it was an introduction to Sun Tzu and the art of strategy. Little did I know that this would spark a lifelong fascination with strategic thinking, samurai history, and Miyamoto Musashi's book "The Book of Five Rings" by.

Fast forward a few decades, and I'm the CISO for the Commonwealth of Pennsylvania, navigating the wild west of early cybersecurity. I couldn't help but notice that while everyone was busy pushing the latest compliance checklists and security gadgets, strategic thinking was taking a nap. It was like showing up to a sword fight with a Swiss Army Knife – lots of tools, but no clue how to use them effectively.

My mind kept drifting back to Musashi and his Five Rings. Why couldn't his principles of combat apply to this new battlefield? So, in 2010, I started a blog to explore this idea, hoping to bridge the gap between ancient wisdom and modern mayhem.

Life, as it often does, threw a few curveballs, and the blog gathered dust. But the concept never left me. Like a persistent virus, it kept replicating in the background until I recently decided to drag it out of quarantine.

What really reignited the fire? I got started on this path through an executive level cybersecurity conference, of all things. I know, I know, those are typically as exciting as watching paint dry. I remember that this one was... different. As speaker after speaker droned on about the latest tools and compliance mandates, reiterating the same old song and dance, nobody talked about strategy. It was all tactics, no vision. Like a general obsessing over the shine on his boots while ignoring the enemy massing on the horizon. Which made me think about conferences I have attended recently. Not much has changed, the focus is on compliance efforts or the latest tools ("now with AI").

That's when it hit me: Musashi's wisdom wasn't just relevant; it was essential. We weren't just fighting code; we were engaged in a war of wits, a battle of minds. And to win that war, we needed strategy, not just checklists.

"The Five Keys" is the result of that kindled spark. It's my translation of Musashi's timeless strategies into actionable principles for cybersecurity. Think of it as a field manual for the modern cyber warrior, blending ancient wisdom with a bit of my own hard-earned cynicism.

In these pages, we'll explore self-knowledge, agility, decisive action, threat awareness, and the importance of process. We'll look at how to apply these "keys" to everything from incident response to risk management and how to cultivate a security mindset as sharp as any finely honed Japanese katana sword.

So, buckle up, and let's embark on this journey together. And try to keep your head about you, ok?

(For anyone interested, I have preserved the original incomplete blogs, exactly as published, in Appendix F)

CYBER STRATEGY MUSINGS

Training in the Way of Cyber Strategy

Welcome

Welcome to my small corner of the Internet where I hope to share my experiences and lessons learned in the art of cyber security strategy. I have chosen to blog anonymously so I may be free to express opinions that may be contrary to employers and peers in my industry. In my world of information assurance, I have often pondered why there are so many companies and consultants promoting the latest and greatest in compliance, controls and tools, without any thought of using strategic thinking in the cyber battlefield where my kind live. My musings always seem to go back to a simple book written a long time ago. The Book of Five Rings by Miyamoto Musashi where the strategy of warfare is laid out in very simple terms.

Samurai history and thought has always been an interest of mine, as well as strategic thinking. Things like USAF Colonel John Boyd's OODA Loop (for Observe, Orient, Decide and Act) concept have always fascinated me. I have always applied these stategies and concepts in my day to day work, but very rarely see any text or program that teaches one how to integrate them into current day cyber security efforts.

I recently attended an executive level conference on cyber security and the lack of speakers and educational tracks with clear strategic value prompted me to actually start my blog!

In subsequent posts I plan on relating how The Book of Five Rings can be used as Five Keys to Successful Cyber Security Strategy. As events allow I will also post real world examples of the strategy in action (with anonymous sources of course).

Comments, musings and links to other strategic sources are always welcome!

Share this:

Facebook   X X

Customize buttons

Posts

April 2025

M	T	W	T	F	S	S
	1	2	3	4	5	6
7	8	9	10	11	12	13
14	15	16	17	18	19	20
21	22	23	24	25	26	27
28	29	30				

« May

Archives

- May 2016
- February 2010
- August 2009
- April 2009
- March 2009

Blog Stats

- 1,156 hits

Recent Posts

- Planes, Trains and Automobiles – Part Deux – Lessons Learned
- Planes, Trains and Automobiles – The return trip from AUSCERT2010
- AUSCERT2010-Day 3
- The Key Of Knowledge (Earth Book) Part 5
- The Key Of Knowledge (Earth Book) Part 4

Sticky Posts

- About the Cyber Stategist
- The Book of Five Keys – Index
- Key of Knowledge
- Welcome

A Note on Artificial Intelligence

"The art of utilizing strength is this: Even strength is to be examined." – Sun Tzu

This book is a testament to the power of combining human insight with artificial intelligence. In a way, it's a bit like Musashi teaming up with a super-computer.

To be more specific, I employed several AI-driven tools. I used the "Deep Research" feature in Gemini to explore cybersecurity strategy more comprehensively than I could have managed alone. Think of it as having a tireless research assistant who can sift through mountains of data and find connections I might otherwise have missed. The results, included in the appendices, helped me refine my thinking and venture down intellectual paths I hadn't considered.

Gemini also proved invaluable in the writing process itself. I often used it as a sounding board, engaging in a sort of iterative dance of prompt development, refinement, and reconsideration. It was like having a writing partner who never gets tired of brainstorming, even if my initial ideas were – shall we say – less than polished.

And yes, even a seasoned CISO like myself needs a little help with grammar now and then. I used Grammarly to help me catch those pesky typos and grammatical errors that tend to creep in when you're wrestling with complex ideas. Consider it my digital editor's red pen.

Of course, no AI – no matter how advanced – can replace the human touch entirely. Ultimately, a human editor reviewed the concepts and content, ensuring that the book retained my voice and, more importantly, made sense.

So, while this book is rooted in my experience and strategic vision, it's also a product of collaboration with some seriously smart machines. It's my hope that this blend of old-school wisdom and new-school technology provides you with a unique and insightful guide to navigating the cyber battlefield.

Introduction: The Way of the Cyber Strategist

"To know ten thousand things, know one well."
– Miyamoto Musashi

The digital world is not a game; it's a battlefield. Forget the romantic notions of hackers in hoodies and the thrill of the chase. We're in a state of constant conflict, a relentless struggle against adversaries who seek to exploit vulnerabilities, steal data, disrupt critical infrastructure, and sow chaos. The weapons have evolved at an alarming pace, far beyond the simple viruses and website defacements of the past. Today, we face sophisticated ransomware attacks that hold entire organizations hostage, zero-day exploits that bypass traditional defenses, and AI-driven attacks that learn and adapt faster than we can react. The stakes are astronomical. Data breaches expose sensitive information, costing companies millions in financial losses and inflicting lasting reputational damage. Critical infrastructure is targeted, threatening essential services and public safety.

In this high-stakes environment, many organizations operate in a state of perpetual reaction. They chase the latest security tools, scramble to meet compliance mandates, and react to the most recent threat alerts. It's akin to a swordsman who only parries – always defending, never attacking – inevitably tiring and succumbing to the opponent.

What's glaringly absent is strategy

A true cyber strategist doesn't just react to threats; they proactively shape their defenses, anticipate future challenges, and align security initiatives with the overarching goals of the organization. Strategy provides the essential framework, the guiding principles, and the long-term vision needed to navigate the complexities of cybersecurity.

My own journey into the world of cybersecurity has been deeply influenced by the strategic wisdom of the past, particularly the timeless insights of the legendary Japanese swordsman Miyamoto Musashi. His masterpiece, "The Book of Five Rings," is far more than a martial arts manual; it's a profound treatise on strategy, applicable to any domain where conflict and competition exist, including the digital battlefield.

Cyber strategy is not about having the shiniest gadgets; it's about understanding your environment, anticipating your adversaries' moves, and making calculated decisions to achieve your objectives. As Musashi says, "Strategy is the craft of the warrior." It's not merely about knowing how to wield a sword (or a security tool), but about understanding when, where, and why to strike. Cyber strategists understand, as Musashi did, that "Perception is strong and sight weak." Relying solely on immediate observations (like threat alerts) without gaining a deeper understanding of the overall situation is a recipe for disaster.

Musashi's relentless emphasis on self-knowledge, adaptability, and understanding opponent strengths and weaknesses resonates deeply with the challenges faced by cybersecurity professionals in today's digital age. He stresses the importance of "knowing the ways of all things," a concept directly applicable to cybersecurity professionals who must understand not only security technologies but also cyber adversaries' evolving tactics and motivations.

Using the Five Keys to Cultivate a Strategic Mindset

"The Book of Five Keys" translates Musashi's enduring principles into actionable strategies for the modern cyber strategist. Along with the influences of the works of Colonel John Boyd, it provides a comprehensive framework for building a robust and resilient cybersecurity posture that moves beyond simply acquiring the latest tools and solutions by cultivating a strategic mindset and embracing a holistic approach to security. This approach is about moving beyond tactical responses to develop a strategic vision that resonates with Musashi's emphasis on "long-range strategy" in warfare.

The Five Keys presented in this book – Self-Knowledge, Agility, Action, Threats, and Process – are directly derived from Musashi's Five Rings, each offering a unique perspective on the art of strategy:

The Ground Book becomes the foundation for Self-Knowledge, emphasizing the critical importance of understanding your environment, your tools, your own capabilities, and your organization's unique risk profile. Just as Musashi stresses the importance of a solid foundation in swordsmanship, so too is self-knowledge the bedrock of effective cybersecurity.

The Water Book inspires Agility, focusing on adaptability, responsiveness, and maintaining balance in the face of the ever-changing tides of cyber threats. Musashi's concept of water adapting to its container mirrors the need for cybersecurity professionals to be flexible and adjust their strategies to new challenges.

The Fire Book fuels Action, urging proactive defense, offensive security, and taking the initiative in the fight against cybercrime. Musashi's emphasis on seizing the moment and attacking decisively translates to the need for proactive cybersecurity threat hunting and rapid incident response.

The Wind Book highlights the importance of understanding Threats, emphasizing the ever-evolving threat landscape, the critical role of the human element in security, and the specific attack vectors that organizations face. Musashi's advice to "know the enemy, know yourself" is particularly relevant in cybersecurity threat intelligence.

The Book of the Void underscores the significance of Process, encompassing the creation of a strong security culture, robust governance, strategic planning that aligns with business objectives, and the essential need to prepare for the inevitable challenges and uncertainties of the future. Musashi's concept of Void as a state of preparedness and awareness resonates with the need for continuous improvement and adaptation in cybersecurity.

Putting the Five Keys to Work

Ultimately, this book is more than just a collection of strategies and tactics. It goes beyond theoretical concepts and abstract principles, providing practical guidance, relevant real-world examples, and actionable strategies that you can implement within your organization. It also addresses the critical need for effective Third-Party Risk Management (TPRM), a growing concern for organizations of all sizes, and examines emerging trends and technologies like the increasing role of AI in cybersecurity, both as a threat and a defense.

It's an invitation to embrace the "Way of the Cyber Strategist," which is a path of continuous learning, constant adaptation, and proactive defense. This approach:

- cultivates a mindset that accepts challenges and actively seeks them out,
- anticipates threats before they materialize, and
- leverages every opportunity to strengthen your organization's security posture and build true resilience.

As Musashi states, "There is nothing outside of yourself that can help you; that kind of knowledge is obtained inside yourself." This underscores the importance of self-reliance and continuous self-improvement in pursuing cybersecurity excellence.

The journey will undoubtedly be challenging. The digital battlefield is in constant flux, and new threats and attack vectors emerge with alarming regularity. But with the right strategy, mindset, and tools, you can not only survive in this challenging environment but truly thrive.

So, let us embark on this journey together. Let us explore the Five Keys, draw wisdom from the strategic insights of the past, and equip ourselves to face the complex and ever-evolving cybersecurity challenges of the future with confidence and resolve.

Key 1: Self-Knowledge (The Ground Book)

"In cybersecurity, the 'one thing' is mindset — tools change, threats evolve, but the strategist endures." — Bob Maley

Understanding Your Environment

In "The Book of Five Rings" Musashi begins with "The Ground Book," emphasizing the foundational principles of strategy. He states, "In strategy, it is important to see distant things as if they were nearby and to take a distanced view of close things." This principle is directly applicable to cybersecurity. To effectively defend our digital terrain, we must first have a deep and comprehensive understanding of our environment. This foundation requires more than a superficial overview of what assets you have. Musashi stresses the importance of "knowing the ways of all things," and in cybersecurity, this translates to a comprehensive awareness of every facet of your technological landscape. The principles call for cultivating a granular understanding of how those assets interact, their inherent vulnerabilities, their importance and criticality to the organization's mission, and the flow of information within your digital ecosystem.

Asset Classification and Management

Just as a warrior must be intimately familiar with their weapons, a cybersecurity professional must have a meticulous and comprehensive understanding of their assets. This involves more than listing assets, rather it requires a detailed and dynamic inventory of all hardware, software, data, and services that comprise the organization's digital ecosystem. This process is about knowing the tools of your trade better than your adversary knows them.

Hardware: Each device has a role to play, and you must know its strengths and weaknesses. This includes everything from the obvious elements like servers, workstations, and network devices (routers, switches, firewalls) to the often-overlooked components such as IoT devices, industrial control systems (ICS), and mobile devices (smartphones, tablets). Beyond simply listing these devices; you must meticulously document their purpose within the organization, their specific configuration settings, the firmware versions being run, the network segments they reside on, and the individuals or teams responsible for their maintenance and security. Think of it as Musashi's deep understanding of the different types of swords – the katana, the wakizashi, the tanto – and their specific applications in combat.

Software: Like a swordsman knowing the sharpness and balance of their blade, knowledge of software interdependencies allows you to maintain a precise understanding of the state of your software, including its potential flaws. This picture encompasses the vast and complex world of operating systems, applications (both commercial off-the-shelf and custom-developed), databases, middleware, and any other software that run or interact with your systems. You need to rigorously track software versions, applied patches, and known vulnerabilities associated with each software component. This granular documentation includes understanding software dependencies, licensing agreements, and the software development lifecycle (SDLC) of any custom applications.

Data: Like a general meticulously protecting their supply lines and resources, you must implement robust measures to protect your data from unauthorized access, modification, or destruction. In today's information-driven world, data is often an organization's most valuable asset. Therefore, you must meticulously identify where data is stored (databases, file servers, cloud storage), how it is accessed (applications, APIs, user interfaces), how it is transmitted (encrypted channels, protocols), and, crucially, its sensitivity level. Is it personally identifiable information (PII) subject to privacy regulations? Does the data include sensitive financial records? Is it intellectual property that gives your organization a competitive edge?

Services: To be complete, your efforts must include an examination of your services enterprise-wide. This category includes a broad range of network services (DNS, DHCP, email), cloud services (SaaS, PaaS, IaaS), web services (APIs), and any other services, both internal and external, that your organization relies upon for its operations. You need to thoroughly understand the dependencies between these services, their criticality to business processes, their security configurations, and the associated service level agreements (SLAs). This also involves understanding the protocols they use, the ports they communicate on, and the authentication and authorization mechanisms they employ.

Effective asset management isn't a static, one-time activity; it's a dynamic and ongoing process that must adapt to the ever-changing nature of the digital landscape. Your environment is constantly in flux, with new devices being added, software being updated, and services being deployed. Therefore, your asset inventory must be meticulously maintained and updated in real time to mirror these changes. This process requires implementing automated discovery and inventory tools, establishing clear and well-defined asset management processes, and assigning accountability to individuals or teams for the management and security of specific asset categories. This effort also involves integrating asset management with other security functions – such as vulnerability management and incident response – to ensure a holistic and coordinated approach to security.

Vulnerability Management and Proactive Testing

Knowing what assets you have is merely the first step in the journey to self-knowledge. You must also delve deeper to understand the underlying inherent weaknesses and potential points of failure. In cybersecurity, these weaknesses are referred to as vulnerabilities, and identifying and addressing them is paramount to maintaining a strong security posture.

Vulnerability Scanning: Automated vulnerability scanning tools play a crucial role in this process by systematically scanning your systems and applications for known vulnerabilities. These tools compare the software versions and configurations of your assets against databases of known vulnerabilities, providing you with a list of potential weaknesses that could be exploited by an attacker. This is akin to a warrior meticulously inspecting their armor for weak points or chinks in the plating, identifying areas that need reinforcement or repair.

Penetration Testing: Penetration testing can reveal vulnerabilities in application logic, configuration errors, and other weaknesses that require human ingenuity to uncover. While vulnerability scanning provides a valuable baseline assessment, penetration testing takes a more active and offensive approach. Penetration testing involves ethical hackers or security professionals simulating real-world attacks against your systems to identify vulnerabilities that automated scanners might miss. This activity includes attempting to exploit vulnerabilities, bypass security controls, and gain unauthorized access to sensitive data or systems. It's like engaging in a practice duel or a simulated battle to test the effectiveness of your defenses under pressure.

Red Teaming: Red teaming represents an even more advanced and comprehensive form of testing that involves a dedicated team of security professionals simulating a sophisticated and persistent real-world attack scenario. Red teaming exercises go beyond simply identifying vulnerabilities; they aim to evaluate the effectiveness of your entire security program. It tests your organization's ability to detect, respond to, and recover from a targeted attack, including evaluating your technical defenses, your people's awareness and response capabilities, and the effectiveness of your security processes and procedures. This exercise is analogous to a full-scale war game or a complex military exercise, testing not only the individual skills of the warriors but also the coordination and effectiveness of the entire army.

Musashi emphasizes the importance of constant practice, meticulous preparation, and understanding the nuances of combat. Proactive security testing, including vulnerability scanning, penetration testing, and red teaming, is your cybersecurity equivalent of this practice and preparation. It provides you with invaluable insights into the strengths and weaknesses of your defenses, allowing you to identify and address vulnerabilities before an attacker can exploit them in a real-world scenario. It's about taking the initiative and proactively seeking out weaknesses rather than waiting for an attacker to find them for you.

Risk Management Principles and Frameworks

A thorough understanding of your assets and their associated vulnerabilities is essential for implementing effective risk management practices. Risk management is a systematic process that involves identifying, assessing, analyzing, and prioritizing potential risks to your organization's assets, operations, and objectives. This assessment lets you make informed decisions about how to allocate your limited security resources to mitigate the most significant threats.

Musashi emphasizes the importance of "evaluating things rightly" and making sound judgments based on a clear understanding of the situation. Risk management is the cybersecurity equivalent of this evaluation process. It's about systematically evaluating threats and vulnerabilities, analyzing potential impacts, and making informed decisions about how to allocate resources, implement security controls, and prioritize security efforts. It's not about eliminating all risk – which is often an impossible goal – it's about managing risk to an acceptable level that aligns with your organization's risk appetite and business objectives.

Risk Identification: The first step in risk management is to identify potential threats and vulnerabilities that could harm your organization. This involves considering a wide range of potential scenarios, including internal threats, external threats, natural disasters, and human error. What potential threats could exploit the vulnerabilities in your systems? What are the potential impacts of those threats if they were to materialize? This step requires a comprehensive understanding of your threat landscape, including the motivations and capabilities of potential attackers.

Risk Assessment: Once you have identified potential risks, you need to assess the likelihood of occurrence and the potential impact if those risks were to occur. This involves analyzing historical data, threat intelligence reports, and other relevant information to estimate the probability of a threat exploiting a vulnerability. You also need to assess the potential consequences of a successful attack, including financial losses, reputational damage, legal liabilities, and operational disruptions.

Risk Prioritization: Not all risks are created equal. Some risks are more likely to occur and some risks have a greater potential impact than others. Therefore, you need to prioritize risks based on the combination of likelihood and impact, so that you can focus your available resources on mitigating the most significant threats. This involves developing a risk matrix or other risk assessment tool to visualize and compare different risks.

Established cybersecurity frameworks, such as the NIST Cybersecurity Framework (NIST CSF) and ISO 27001, provide valuable guidance and best practices for designing and implementing effective risk management programs. These frameworks provide a structured and methodical approach to assessing and addressing threats, similar to Musashi's structured approach to strategy and combat. They offer a common language and set of standards for organizations to use when managing their cybersecurity risks.

By diligently and comprehensively understanding your environment through meticulous asset classification and management, proactive vulnerability management and testing, and the application of sound risk management principles and frameworks, you are able to lay a solid and unshakeable groundwork for a robust and resilient cybersecurity posture.

This self-knowledge is the fundamental bedrock upon which all other security efforts are built. Without it, you're essentially fighting blindfolded, lacking the situational awareness and understanding necessary to effectively defend your organization against the ever-present and evolving threats of the digital world. As in warfare, in the world of cybersecurity, fighting blindfolded is a fight you are almost certainly going to lose.

Tools of the Trade

SIEM Systems and Log Analysis

Musashi emphasizes the critical importance of knowing your tools and mastering their application. He advises, "Familiarize yourself with the tools of your craft, for a skilled craftsman is only as good as the tools they wield." In the context of cybersecurity, this instruction translates to a deep and comprehensive understanding of the technologies, systems, and techniques that empower us to effectively monitor, analyze, and defend our digital environments. SIEM systems and log analysis are not merely supplementary components but fundamental tools in a cybersecurity professional's arsenal, representing the eyes and ears of a modern digital defense.

SIEM Systems: The Central Nervous System of Security Monitoring

A Security Information and Event Management (SIEM) system functions as an organization's central nervous system, operational hub, and strategic control center for security monitoring and incident response capabilities. This system is the critical juncture where data from a multitude of diverse sources converges, providing security teams with a holistic and comprehensive view of security activity across the entire organization.

Functionality: SIEM systems perform a range of critical functions that are essential for maintaining a strong security posture:

→ **Log Collection and Management:** SIEM systems are designed to gather and aggregate logs and event data from a vast array of sources, including servers, workstations, network devices (routers, switches, firewalls, intrusion detection/prevention systems), applications (web applications, databases, custom applications), security devices (endpoint detection and response (EDR) solutions, vulnerability scanners, identity and access management (IAM) systems), and cloud services. This process involves not only collecting the data but also normalizing it, standardizing it, and storing it in a centralized repository for efficient analysis and retrieval. This is akin to a warrior meticulously gathering intelligence and reports from all corners of the battlefield, ensuring that no piece of information, no matter how small, is overlooked or lost.

→ **Event Correlation and Analysis:** One of the core strengths of a SIEM system lies in its ability to analyze and correlate security events from different sources to identify complex patterns, detect anomalies, and uncover suspicious activities that might indicate a security incident or an ongoing attack. This goes beyond looking for isolated events by connecting the dots, identifying relationships, and understanding the context of security events. This is like a seasoned general analyzing reports from different scouts, intelligence sources, and battlefield observations to understand the enemy's overall strategy, anticipate their next move, and identify hidden threats that might not be apparent at first glance.

→ **Alerting and Notification:** SIEM systems are equipped with robust alerting and notification capabilities that enable security teams to respond quickly and effectively to security incidents. Based on predefined rules, correlation logic, or anomaly detection algorithms, when suspicious activity is detected, the SIEM system automatically generates alerts and notifications, notifying security personnel of the potential threat. These alerts can be prioritized based on severity and criticality, allowing security teams to focus on the most urgent issues. This is like an early warning bell or alarm system that signals an impending attack or breach of security, providing defenders with the time and opportunity to react and mitigate the damage.

→ **Reporting and Compliance:** SIEM systems play a crucial role in providing comprehensive reports on security events, trends, and compliance status. These reports can be used for a variety of purposes, including incident analysis, security audits, compliance reporting (e.g., GDPR, HIPAA, PCI DSS), and performance monitoring. SIEM systems can generate customized reports based on specific requirements, providing valuable insights into the organization's security posture and helping to identify areas for improvement. This is like a detailed post-battle report, meticulously analyzing what happened during the engagement, identifying the successes and failures, and providing valuable insights for future strategies and training.

Examples: The SIEM landscape is diverse, with a range of solutions catering to different organizational needs and budgets. Some of the popular SIEM solutions include:

→ **Splunk:** A widely recognized and highly regarded platform known for its powerful search and analysis capabilities, its flexibility, and its ability to handle massive volumes of data. Splunk is often used by large enterprises with complex security requirements.

→ **IBM Security QRadar SIEM:** A comprehensive and enterprise-grade SIEM solution that offers advanced analytics, robust threat intelligence integration, and strong incident response capabilities. QRadar is known for its ability to provide deep insights into network activity and identify sophisticated threats.

→ **Microsoft Sentinel:** A cloud-native and scalable SIEM solution that leverages the power of AI and machine learning to enhance threat detection, automate response, and provide proactive security insights. Sentinel is particularly well-suited for organizations that heavily rely on Microsoft's cloud services.

→ Other Notable SIEM Solutions: Other notable SIEM solutions include Sumo Logic, McAfee Enterprise Security Manager, and Elastic Security – each with its own unique strengths and features.

AI's Transformative Impact: Artificial intelligence (AI) is having a profound and transformative impact on SIEM systems, revolutionizing the way organizations approach security monitoring, threat detection, and incident response. AI is not just an add-on feature; it's fundamentally changing the capabilities and effectiveness of SIEM systems.

→ **Enhanced Threat Detection:** AI algorithms, including machine learning models, are capable of analyzing vast amounts of data from diverse sources and identifying subtle anomalies, deviations from normal behavior, and complex patterns that human analysts might easily miss. This significantly enhances the ability of SIEM systems to detect sophisticated and evasive threats, such as zero-day exploits, insider threats, and advanced persistent threats (APTs). This is like having a super-powered scout or a highly trained intelligence officer who possesses the ability to spot hidden enemies, detect subtle clues, and anticipate threats that would be invisible to the untrained eye.

→ **Automated Response and Orchestration:** AI can be leveraged to automate certain security tasks and orchestrate incident response actions, reducing the time it takes to respond to security incidents, thereby minimizing the potential impact of attacks. AI-integrated SIEM systems can trigger pre-defined automatic response actions, such as isolating infected devices, blocking malicious IP addresses, disabling compromised user accounts, or initiating automated investigations. This is like having highly trained and disciplined soldiers who can automatically defend against minor attacks, execute pre-defined battle plans, and take immediate action to contain the damage and protect the overall force.

→ Predictive Security and Threat Forecasting: AI can utilize historical data, threat intelligence feeds, and behavioral analysis to predict potential future attacks, enabling security teams to proactively strengthen their defenses, anticipate emerging threats, and take preventative measures to mitigate risks before they materialize. This mirrors the actions of a highly skilled strategist or seasoned military commander who possesses the ability to anticipate the enemy's next move, predict their attack vectors, and develop proactive countermeasures to neutralize the threat before it can cause harm.

Log Analysis: Deciphering the Digital Footprints

Logs are the detailed and comprehensive digital footprints of system and user activity, recording events, actions, and transactions that occur within a computer system, network, or application. Analyzing these logs is of paramount importance for gaining a deep understanding of what's happening within your digital environment, detecting security incidents, investigating suspicious activity, and troubleshooting technical problems. Log analysis is the cornerstone of effective security monitoring and incident response.

Importance: Log analysis provides invaluable insights and serves several critical purposes in cybersecurity:

→ **Incident Investigation and Forensics:** Logs are essential for conducting thorough incident investigations and performing digital forensics. In the aftermath of a security incident, logs provide the crucial evidence needed to understand the scope and impact of the attack, determine the root cause, identify the attackers, and reconstruct the sequence of events. This is like a skilled detective meticulously piecing together clues, analyzing forensic evidence, and reconstructing the crime scene to solve a complex case and bring the perpetrators to justice.

→ **Threat Hunting and Proactive Detection:** Proactive log analysis is a powerful technique for identifying hidden threats, detecting suspicious activity, and uncovering malicious actors that might be lurking within the environment, evading traditional security controls. By actively searching for anomalies, investigating suspicious patterns, and correlating events across different log sources, security teams can proactively identify and neutralize threats before they cause significant damage. This is like a skilled hunter meticulously tracking their prey, searching for subtle signs, and anticipating their movements to capture them before they can escape.

→ **Performance Monitoring and Troubleshooting:** Logs also provide valuable insights into system performance, application behavior, and network activity. Analyzing logs can help identify performance bottlenecks, diagnose technical errors, troubleshoot application problems, and optimize system performance. This is like a skilled mechanic diagnosing a car problem, using specialized tools and analyzing engine data to identify the root cause of the issue and implement effective repairs.

→ **Techniques:** Log analysis involves a variety of techniques, ranging from manual inspection to sophisticated automated analysis.

Manual Analysis: Security analysts can manually review and inspect log files to identify suspicious activity, investigate security incidents, or troubleshoot technical problems. This approach is often used for smaller datasets or for specific investigations where a human analyst's expertise and intuition are required. This analysis follows the model used by a dedicated soldier carefully inspecting the terrain, searching for subtle signs of enemy activity, or meticulously examining a piece of evidence.

→ **Automated Analysis:** A wide range of tools and technologies can automate log analysis, using rules, filters, regular expressions, and other techniques to identify patterns, detect anomalies, and trigger alerts based on predefined criteria. Automated log analysis is essential for processing and analyzing large volumes of log data efficiently and effectively. This process is using a powerful machine or a specialized tool to sort through large amounts of data, identify specific patterns, and automate repetitive tasks, thereby allowing human analysts to focus on more complex and critical issues.

→ **Machine Learning and AI-Powered Analysis:** Machine learning algorithms and AI-powered log analysis tools are increasingly being used to enhance log analysis capabilities, providing more sophisticated and accurate threat detection, anomaly detection, and behavioral analysis. Machine learning algorithms can learn from historical data, identify subtle anomalies that rule-based systems might miss, and adapt to evolving threat patterns. This is like having a highly trained and experienced analyst who possesses the ability to spot hidden patterns, detect subtle clues, and anticipate threats that would be invisible to the untrained eye, leveraging advanced techniques and sophisticated tools to enhance their analytical capabilities.

AI's Transformative Impact: AI is transforming the field of log analysis, providing security teams with more powerful, efficient, and effective tools for detecting threats, investigating incidents, and gaining deeper insights into their digital environments. AI is not just an enhancement; it's a game-changer in log analysis.

→ **Enhanced Anomaly Detection:** AI algorithms, particularly machine learning models, excel at detecting anomalies in log data that might indicate a security incident or malicious activity. AI-powered log analysis tools can learn from normal behavior patterns, identify deviations from the baseline, and flag suspicious events that might not be detected by traditional rule-based systems. This is like having a highly sensitive and intelligent guard dog that can sense intruders, detect unusual behavior, and alert security personnel to potential threats.

→ **Behavioral Analysis and User Activity Monitoring:** AI can analyze user and system behavior patterns, identify suspicious or anomalous actions, and detect insider threats or compromised accounts. AI-powered log analysis tools can monitor user activity, track access patterns, and detect deviations from normal behavior, providing valuable insights into potential security risks. This is like a skilled detective profiling a suspect, analyzing their behavior patterns, and identifying suspicious actions that might indicate criminal intent.

→ **Threat Intelligence Integration and Contextual Enrichment:** AI can seamlessly integrate threat intelligence feeds with log analysis, providing real-time context and enriching log data with information about known threats, malicious IP addresses, and other indicators of compromise (IOCs). This allows security teams to quickly identify and prioritize known threats, reduce false positives, and make more informed decisions about security incidents. This is like having access to a comprehensive and up-to-date database of known criminals, their tactics, and their associates that allows security personnel to identify and neutralize known threats quickly.

Musashi emphasizes the importance of keen observation, meticulous attention to detail, and a deep understanding of your opponent's strengths and weaknesses. SIEM systems and log analysis provide the essential tools for observation, monitoring, and intelligence gathering in the digital realm. AI significantly enhances these tools, providing greater visibility, deeper insights, and enabling more effective threat detection and incident response.

Mastering SIEM systems and log analysis, leveraging the power of AI, and embracing a proactive and strategic approach to security monitoring enables cybersecurity professionals to gain a deeper and more comprehensive understanding of their digital environments, detect threats with greater accuracy and speed, respond to incidents more effectively, and ultimately build a stronger foundation of self-knowledge and achieve mastery in the Way of the Cyber Strategist. This mastery is crucial for navigating the complexities and challenges of the modern cybersecurity landscape and effectively defending organizations against the ever-evolving threats of the digital world.

Penetration Testing Tools and Techniques

Musashi emphasizes the importance of not only knowing your tools but also achieving mastery in their application through rigorous practice and a deep understanding of their capabilities and limitations. He states, "You must train in all of the different styles of combat, for each has its strengths and weaknesses, and you must know them all to be truly effective." In cybersecurity, penetration testing embodies this principle by simulating a wide range of attack scenarios to rigorously test the effectiveness of your defenses and identify potential vulnerabilities before malicious actors can exploit them. This is an ongoing process for proactively seeking out weaknesses, understanding the nuances of your security posture, and continuously refining your defensive strategies.

The Art of Ethical Hacking

Penetration testing is a crucial and indispensable aspect of achieving self-knowledge and maintaining a strong security posture in cybersecurity. It involves ethical hackers or other highly skilled security professionals simulating real-world attacks against your systems, networks, applications, and even your people to identify vulnerabilities, weaknesses, and security flaws that could be exploited by malicious actors. It's not just about running automated tools; it's about applying critical thinking, creativity, and a deep understanding of attack methodologies to put your defenses to the ultimate test and gain a comprehensive understanding of their strengths and limitations.

The   Arsenal   of   the   Ethical   Hacker

Penetration testers utilize a diverse and ever-evolving arsenal of tools to simulate various attack vectors and uncover vulnerabilities. These tools range from network scanners and vulnerability assessment tools to specialized web application scanners, exploitation frameworks, and password cracking utilities. Each tool serves a specific purpose, and a skilled penetration tester knows how to select and use the right tool for the job.

Network Scanners: These are foundational tools used to discover hosts, identify services, and map the network topology of a target environment. They play a crucial role in the reconnaissance and information gathering phase of a penetration test, providing valuable insights into the target's network infrastructure.

→ **Nmap:** Arguably the most powerful and versatile network scanner available, Nmap (Network Mapper) is an indispensable tool for penetration testers. It's used for a wide range of tasks, including host discovery (identifying active hosts on a network), port scanning (determining which ports are open and listening on a host), service identification (identifying the services running on those ports), operating system detection, and even basic vulnerability scanning. Nmap's flexibility, scripting capabilities, and extensive feature set make it a cornerstone of network penetration testing. You can find more information at nmap.org.

→ **Nessus:** While Nmap focuses on network discovery and basic service identification, Nessus takes it a step further by providing comprehensive vulnerability scanning capabilities. It can identify vulnerabilities in operating systems, applications, databases, and network devices, providing detailed reports on potential security flaws. Nessus is a commercial tool but offers a free "Home" version for personal use. You can explore its features at tenable.com/nessus.

Vulnerability Scanners: These tools go beyond simple network scanning and delve deeper into identifying specific vulnerabilities in applications, systems, and network devices. They often utilize databases of known vulnerabilities to compare the target's configuration against known security flaws.

→ **OpenVAS:** A powerful and versatile free and open-source vulnerability scanner, OpenVAS (Open Vulnerability Assessment System) provides a comprehensive suite of tools for vulnerability scanning and management. It can identify a wide range of vulnerabilities in web applications, servers, and network devices, making it a valuable alternative to commercial vulnerability scanners. You can learn more at openvas.org.

→ **Qualys:** A leading cloud-based vulnerability management platform, Qualys provides a comprehensive suite of services for vulnerability scanning, threat detection, and compliance management. It offers a wide range of scanning capabilities, including web application scanning, cloud security assessment, and endpoint vulnerability detection. Qualys is a commercial platform designed for enterprise-level vulnerability management. You can find more information at qualys.com.

Web Application Scanners: With the increasing reliance on web applications, these specialized tools are essential for identifying vulnerabilities in web interfaces, APIs, and web servers. They are designed to detect common web application flaws, such as SQL injection, cross-site scripting (XSS), and authentication bypass vulnerabilities.

→ **Burp Suite:** A highly popular and widely used web application testing tool, Burp Suite is a comprehensive platform that includes a proxy (for intercepting and modifying web traffic), a scanner (for automated vulnerability detection), and an intruder (for customized attack automation). Burp Suite is considered an industry-standard tool for web application penetration testing. You can explore its capabilities at portswigger.net/burp.

→ **OWASP ZAP:** The OWASP Zed Attack Proxy (ZAP) is a free and open-source web application security scanner that provides a user-friendly interface and a wide range of features for web application penetration testing. It's a valuable tool for both beginners and experienced penetration testers. You can learn more at owasp.org/zaproxy.

Exploitation Frameworks: These powerful frameworks provide a collection of pre-built exploits, payloads, and tools that can be used to test identified vulnerabilities and gain access to target systems. They streamline the exploitation process and provide a standardized platform for penetration testers.

→ **Metasploit:** Developed by Rapid7, Metasploit is the most widely used and recognized exploitation framework in the world. It includes a vast database of exploits for various vulnerabilities, as well as tools for payload generation, encoding, and delivery. Metasploit is an essential tool for penetration testers, providing a powerful platform for testing and validating vulnerabilities. You can find more information at rapid7.com/metasploit.

Password Cracking Tools: These tools are used to test the strength of passwords and identify weak or easily guessable credentials. They employ various techniques, such as dictionary attacks, brute-force attacks, and rainbow table lookups, to attempt to crack password hashes.

→ **John the Ripper:** A highly versatile and widely used password cracking tool, John the Ripper supports a wide range of hashing algorithms and can be used to crack passwords on various platforms. It's known for its flexibility and its ability to be customized for specific password cracking tasks. You can learn more at openwall.info/john.

→ **Hashcat:** A powerful and extremely fast password cracking tool, Hashcat leverages the processing power of GPUs (Graphics Processing Units) to accelerate the password cracking process. It supports a wide range of hashing algorithms and is known for its speed and efficiency. You can find more information at hashcat.net.

Rainbow Tables: The Illusion of Simplicity in Cracking the Cipher

In the ever-evolving arms race of cybersecurity, few tools better illustrate the cunning efficiency of attackers than the rainbow table. At its core, a rainbow table is a precomputed set of hashes derived from plaintext passwords using a specific hashing algorithm. Rather than brute-forcing every possible combination in real-time, an attacker can use this table to reverse cryptographic hashes by simply looking up the match – much like a swordsman who already knows exactly where the armor is weakest. The danger lies in the speed; what once took hours or days can be executed in seconds. However, rainbow tables are not invincible. They rely on the absence of randomness – no salting, no uniqueness. As a cyber strategist, one must understand that defending against rainbow tables isn't about building an unbreakable wall; it's about introducing unpredictability. Properly implemented salting renders rainbow tables obsolete, turning the attacker's shortcut into a dead end. Like Musashi's advice to never meet the enemy head-on when misdirection is possible, smart defenders use entropy and foresight to neutralize brute efficiency.

The Art of Deception and Exploitation

Penetration testing is not just about running tools; it's a strategic process involving technical skills, critical thinking, and a deep understanding of attack methodologies. It's about thinking like an attacker to identify potential weaknesses and exploit them in a controlled and ethical manner.

Reconnaissance: Gathering Intelligence and Mapping the Target. Reconnaissance is the initial and crucial phase of any penetration test. It involves gathering as much information as possible about the target organization, its systems, and its people. This phase is analogous to a military scout gathering intelligence about the enemy's positions, strengths, and weaknesses.

→ **Open-Source Intelligence (OSINT) Gathering:** This process involves collecting information from publicly available sources, such as websites, social media profiles, public records, and search engines. OSINT can provide valuable insights into the target's technology stack, employee information, and potential vulnerabilities.

→ **Network Scanning:** This type of scan involves using network scanning tools, such as Nmap, to identify active hosts, open ports, and running services on the target network. It provides a map of the target's network infrastructure and helps identify potential attack vectors.

→ **Social Engineering:** While not always included in every penetration test, social engineering involves gathering information through human interaction, such as phishing emails, phone calls, or in-person interactions. This can be used to gather sensitive information, such as login credentials or internal procedures.

Scanning: Identifying Vulnerabilities and Weaknesses

The scanning phase involves using specialized tools, such as vulnerability scanners and web application scanners, to identify potential weaknesses in the target systems and applications. This phase builds upon the information gathered during reconnaissance and focuses on identifying specific security flaws that could be exploited.

→ **Vulnerability Scanning:** This involves using automated vulnerability scanners, such as Nessus or OpenVAS, to identify known vulnerabilities in operating systems, applications, and network devices. These scanners compare the target's configuration against databases of known vulnerabilities and provide reports on potential security flaws.

→ **Port Scanning:** This involves using network scanning tools, such as Nmap, to identify open ports and services running on the target systems. Open ports can be potential entry points for attackers, and identifying the services running on those ports can provide clues about potential vulnerabilities.

Exploitation: Gaining Access and Demonstrating Impact

The exploitation phase involves attempting to exploit identified vulnerabilities to gain unauthorized access to target systems or data. This is where the penetration tester demonstrates the real-world impact of the vulnerabilities and validates the effectiveness of the target's security controls.

Exploiting Known Vulnerabilities: This involves using pre-built exploits or developing custom exploits to take advantage of identified vulnerabilities. Exploits are often available for common vulnerabilities, and penetration testers use them to gain access to systems or execute arbitrary code.

→ **Password Cracking:** This involves using password cracking tools, such as John the Ripper or Hashcat, to attempt to crack passwords and gain access to user accounts or systems. Password cracking can be used to test the strength of passwords and identify weak or easily guessable credentials.

→ **Credential Stuffing:** When the Enemy Reuses Your Own Weapons: Credential stuffing is the digital equivalent of an enemy stealing your sword and turning it against you. It exploits the all-too-human habit of reusing usernames and passwords across multiple systems. Attackers use automated tools to flood login portals with stolen credentials—often harvested from previous data breaches— hoping some will unlock new doors. Unlike brute force, this tactic doesn't guess; it assumes you've already been compromised elsewhere. The real danger is scale: thousands of attempts per second, often undetected by traditional defenses. Defeating this tactic requires layered strategy – multi-factor authentication, anomaly detection, and user education. As Musashi taught, you must assume your opponent is clever, relentless, and already knows your habits. Complacency is the crack they exploit.

→ **Social Engineering:** In some cases, social engineering techniques can be used to manipulate individuals into divulging sensitive information or granting access to systems. This can involve phishing emails, pretexting, or other forms of deception.

Post-Exploitation: Maintaining Access and Expanding Influence

The post-exploitation phase involves actions taken after gaining initial access to a system or network. The goal is to maintain access, establish a persistent presence, and potentially move laterally to other systems on the network.

→ **Maintaining Access:** This involves establishing a backdoor or other mechanism to ensure persistent access to the compromised system, even if the initial entry point is patched.

→ **Lateral Movement:** This involves moving from the initially compromised system to other systems on the network, often with the goal of gaining access to more sensitive data or higher-privilege accounts.

→ Data Exfiltration: This involves stealing sensitive data from the compromised systems or network, such as confidential documents, customer data, and other valuable information.

Reporting: Documenting Findings and Providing Recommendations

The final and crucial phase of a penetration test is reporting. This phase involves documenting all the findings of the penetration test in a clear, concise, and comprehensive report.

The report should include:

→ **Executive Summary:** A high-level overview of the findings, including the most significant vulnerabilities and their potential impact.

→ **Detailed Findings:** A detailed description of each vulnerability, including its location, impact, and the steps taken to exploit it.

→ **Proof of Concept:** Evidence demonstrating the successful exploitation of each vulnerability, such as screenshots or command outputs.

→ **Risk Assessment:** An assessment of the risk posed by each vulnerability, considering its likelihood and potential impact.

→ **Recommendations for Remediation:** Specific and actionable recommendations for fixing the identified vulnerabilities and improving the overall security posture.

Musashi emphasizes the importance of adaptability, resourcefulness, and the ability to apply different techniques depending on the situation. Penetration testing demands a similar approach. This endeavor is not just about blindly running tools; it's about understanding the target environment, choosing the appropriate techniques, adapting your approach as needed, and thinking creatively to overcome challenges.

A skilled penetration tester is a master of their craft, combining technical expertise with strategic thinking to effectively identify and exploit vulnerabilities. Cybersecurity professionals can gain invaluable self-knowledge about the strengths and weaknesses of their defenses by mastering penetration testing tools and techniques. This knowledge empowers them to strengthen their security posture, prioritize remediation efforts effectively, and proactively protect their organizations from real-world attacks. By embracing the mindset of an attacker, they become a more effective defender and ultimately achieve mastery in the Way of the Cyber Strategist.

Threat Intelligence Platforms: Knowing Your Adversary in the Digital Age

Musashi emphasizes the paramount importance of understanding your enemy, their tactics, their motivations, and their capabilities. He writes, "Know the enemy, know yourself, and victory is not in doubt. Know yourself, know your enemy, and you will not be defeated in a hundred battles. Know yourself but not your enemy, and for every victory gained you will also suffer a defeat. Know neither yourself nor your enemy, and you are sure to be defeated in every battle."

In the context of cybersecurity, threat intelligence platforms provide the essential tools and information necessary to "know the enemy" in the digital realm by effectively gathering, meticulously analyzing, and rapidly disseminating actionable intelligence about current and potential threats, threat actors, and their evolving tactics, techniques, and procedures (TTPs).

SIDE NOTE : Throughout "The Book of Five Keys," whenever I use the term "actionable intelligence" it is employed deliberately, aligning with the nuanced understanding detailed in the Appendix G study, rather than the often-simplified usage prevalent in cybersecurity marketing. As the study [in Appendix G] clarifies, transforming raw data or alerts into intelligence that truly supports effective decision-making, especially beyond immediate technical responses, requires significant human analysis, contextualization specific to the organization's environment and risks, and the development of tailored recommendations.[1] Therefore, references to actionable intelligence within this book imply this deeper level of processed, contextualized, and human-validated insight, distinct from vendor claims that may overemphasize automation and minimize the critical role of human judgment.

Empowering Proactive Defense Through Actionable Intelligence

Threat intelligence platforms (TIPs) are not merely passive repositories of data; they are dynamic and essential tools that empower organizations to gain deep self-knowledge about the external threat landscape, anticipate potential attacks, proactively defend their digital assets, and make informed security decisions. They provide a centralized and integrated platform for collecting, meticulously analyzing, enriching, and efficiently disseminating threat information, enabling organizations to move beyond reactive security measures and adopt a proactive and intelligence-driven approach to cybersecurity.

Functionality: The Core Capabilities of Threat Intelligence Platforms (TIPs)

Threat intelligence platforms perform a range of crucial functions that are essential for effective threat management and proactive defense:

Threat Data Aggregation and Collection: TIPs act as central hubs for gathering and aggregating threat data from a multitude of diverse sources, both internal and external. This process involves collecting raw threat data and intelligence from various sources, including:

→ **Open-Source Intelligence (OSINT):** These sources include publicly available information from news articles, blog posts, social media, security research publications, and threat intelligence reports. OSINT provides valuable insights into emerging threats, threat actor activity, and vulnerabilities.

→ **Commercial Threat Intelligence Feeds:** Subscription-based services provide curated and enriched threat intelligence data through reputable vendors. Commercial feeds often provide more timely, accurate, and actionable intelligence than OSINT sources.

→ **Industry Information-Sharing Groups and Communities:** Organizations often participate in industry-specific information-sharing groups and communities, such as Information Sharing and Analysis Centers (ISACs), to exchange threat intelligence with their peers. This collaborative approach allows organizations to benefit from the collective knowledge and experience of the community.

→ **Internal Sources:** TIPs also collect threat data from internal sources, such as security information and event management (SIEM) systems, intrusion detection systems (IDS), firewalls, and endpoint detection and response (EDR) solutions. This internal data provides valuable context and helps organizations understand how threats are impacting their specific environment.

This comprehensive threat data collection is akin to a skilled general gathering intelligence from a diverse network of spies, scouts, allies, and internal reconnaissance teams, ensuring that no potential source of information is overlooked and that a complete picture of the threat landscape is obtained.

Threat Data Analysis and Enrichment: TIPs go beyond simply collecting threat data; they also analyze and enrich the data to provide valuable context, identify trends, and uncover emerging threats. This process involves a range of sophisticated techniques, including:

→ **Machine Learning (ML):** ML algorithms are used to analyze large datasets, identify patterns, detect anomalies, and predict future threat activity. ML can automate the analysis process, improve accuracy, and identify threats that might be missed by human analysts.

→ **Natural Language Processing (NLP):** NLP techniques are used to analyze unstructured data, such as text from blog posts or social media, to extract relevant threat intelligence. NLP can help automate the analysis of OSINT sources and identify emerging trends.

→ **Link Analysis:** Link analysis techniques are used to identify relationships between different threat indicators, such as IP addresses, domain names, and file hashes. This can help uncover complex attack campaigns and identify the infrastructure used by threat actors.

→ Contextual Enrichment: TIPs enrich threat data with contextual information, such as the reputation of IP addresses, the geolocation of threat actors, and the known vulnerabilities associated with specific software. This enrichment provides valuable context and helps security teams prioritize and respond to threats effectively.

This threat data analysis and enrichment process is analogous to an intelligence analyst meticulously piecing together information from various sources, identifying patterns and relationships, and developing a comprehensive understanding of the enemy's plans, capabilities, and intentions.

Threat Intelligence Dissemination and Integration: Threat Intelligence Platforms are designed to efficiently disseminate threat intelligence to various security tools and teams within the organization, enabling them to leverage this intelligence for proactive defense and improved threat detection. This dissemination process involves:

→ **SIEM Integration:** TIPs integrate with security information and event management (SIEM) systems to provide real-time threat intelligence that can be used to enhance threat detection, prioritize alerts, and improve incident response.

→ **IDS/IPS Integration:** TIPs also integrate with intrusion detection systems (IDS) and intrusion prevention systems (IPS) to provide updated threat signatures and improve the ability to detect and block malicious traffic.

→ **SOAR Integration:** Integration with security orchestration, automation, and response (SOAR) platforms enables organizations to automate threat response actions based on threat intelligence.

→ **Threat Intelligence Sharing Platforms (MISP):** TIPs can also share threat intelligence with other organizations and industry groups through platforms like MISP (Malware Information Sharing Platform), fostering collaboration and improving collective defense.

This threat intelligence dissemination and integration process is akin to a military commander sharing critical intelligence with their troops and key allies, ensuring that everyone on the front lines has the information they need to defend themselves and achieve their objectives.

Threat Intelligence Sharing and Collaboration: TIPs play a crucial role in facilitating the sharing of threat intelligence with other organizations, industry groups, and the broader security community. This collaborative approach allows organizations to leverage the collective knowledge and expertise of the community, improve their situational awareness, and enhance their ability to defend against common threats. This is analogous to allies sharing vital information and coordinating their efforts to defeat a common enemy, recognizing that collaboration is essential for achieving collective security.

Benefits of Threat Intelligence: Empowering Proactive and Informed Security Decisions

Leveraging threat intelligence provides organizations with a wide range of benefits, enabling them to enhance their security posture, improve their threat detection and response capabilities, and make more informed security decisions:

→ **Proactive Defense and Prevention:** Threat intelligence empowers organizations to proactively identify and mitigate potential threats before they can cause significant damage. By understanding the tactics, techniques, and procedures (TTPs) of threat actors, organizations can implement proactive security measures to prevent attacks, such as patching vulnerabilities, hardening systems, and implementing stronger security controls. This is akin to preparing for an attack before it happens, anticipating the enemy's moves, and taking proactive steps to neutralize the threat before it can materialize.

→ **Improved Threat Detection and Response:** Threat intelligence enhances the ability of security tools and security teams to detect and respond to cyberattacks more effectively. By providing real-time information about known threats, threat indicators, and malicious activity, threat intelligence enables security tools to identify and block malicious traffic, detect suspicious behavior, and prioritize security alerts. This is like providing your soldiers with better weapons, improved training, and real-time battlefield intelligence, significantly enhancing their ability to defend themselves and defeat the enemy.

→ **Prioritized Response and Efficient Resource Allocation:** Threat intelligence helps organizations prioritize their security efforts and allocate their limited resources more efficiently by focusing on the most relevant and critical threats. By understanding the likelihood and potential impact of different threats, organizations can prioritize remediation efforts, allocate security budgets effectively, and focus their attention on the areas that pose the greatest risk. This is like a military commander focusing their forces on the most important objectives, allocating resources strategically, and prioritizing their efforts to achieve the greatest impact.

→ **Enhanced Situational Awareness and Risk Management:** Threat intelligence provides organizations with a deeper and more comprehensive understanding of the evolving threat landscape, their specific risks, and their overall security posture. By staying informed about emerging threats, threat actor activity, and industry trends, organizations can make more informed decisions about their security strategy, risk management, and compliance efforts. This is like having a clear and comprehensive view of the battlefield, allowing a general to assess the situation, understand the risks, and make informed decisions about their strategy and tactics.

Black Kite Risk Intelligence: Specializing in Third-Party Cyber Risk Management

Side Note: I am currently the Chief Security Officer at Black Kite, and my team uses the platform. I include the information here as I believe it is extremely valuable.

Black Kite (https://blackkite.com/) is a company that provides a specialized risk intelligence platform with a strong focus on third-party risk management. Their platform offers comprehensive and actionable intelligence on the cybersecurity posture of third-party vendors, suppliers, and partners, enabling organizations to effectively manage the risks associated with their extended supply chain and vendor ecosystem.

Third-Party Risk Focus and Supply Chain Security: Black Kite specializes in providing organizations with unparalleled visibility into the cybersecurity risks associated with their third-party vendors, suppliers, and partners. In today's interconnected digital world, organizations increasingly rely on a complex network of third-party relationships for critical services, data processing, and business operations. This reliance creates significant third-party risks, as a security breach at a third-party vendor can have cascading effects on the organization's security and operations. As my LinkedIn article "Third-Party Risk Management: Navigating the Blind Spot" aptly mentions, "In 2023, there were 63 attacks on vendors: from those 63 attacks, 298 data breaches occurred across impacted companies," highlighting the critical importance of effective third-party risk management. Black Kite's platform provides the tools and intelligence necessary to manage these risks effectively.

Cyber Risk Ratings and Benchmarking: Black Kite provides comprehensive and data-driven cyber risk ratings for organizations, enabling companies to quickly and accurately assess the security posture of their third-party vendors and partners. These ratings are based on a wide range of data sources, including technical assessments, open-source intelligence, and proprietary analysis techniques. Black Kite's cyber risk ratings provide a standardized and objective way to compare the security performance of different third parties and benchmark their security posture against industry peers.

Continuous Monitoring and Alerting: Black Kite offers continuous monitoring of third-party cyber risk, providing real-time alerts and notifications when significant changes occur in the security posture of third-party vendors. This continuous monitoring enables organizations to stay informed about potential risks, detect emerging threats, and take proactive action to mitigate them before they can cause harm. This proactive approach is essential for managing the dynamic and evolving nature of third-party cyber risks.

Actionable Risk Intelligence and Remediation Guidance: Black Kite goes beyond simply providing risk ratings; they also provide actionable insights, recommendations, and remediation guidance to help organizations effectively mitigate third-party cyber risks. This can include specific guidance on remediation steps, best practices for improving third-party security, and tools for communicating and collaborating with vendors to address identified risks. Black Kite's focus on actionable insights empowers organizations to take concrete steps to improve their third-party security and reduce their overall risk exposure.

Musashi emphasizes the importance of gathering comprehensive information, understanding your opponent's strengths and weaknesses, and anticipating their moves. Threat intelligence platforms, particularly those like Black Kite that specialize in third-party risk intelligence, provide the essential tools and information necessary to gain this critical understanding in the complex and interconnected digital realm. By leveraging threat intelligence, organizations can significantly enhance their self-knowledge about the external threat landscape, proactively defend against cyberattacks, effectively manage third-party risks, and ultimately improve their overall security posture and resilience in the face of evolving cyber threats.

Security Automation Tools - Enhancing Efficiency, Effectiveness, and Resilience

Musashi emphasizes the paramount importance of efficiency, effectiveness, and adaptability in combat. He advises, "Do nothing which is of no use. Efficiency is the key to victory." In cybersecurity, security automation tools embody this principle by automating repetitive tasks, streamlining security operations, enhancing incident response capabilities, and enabling security teams to respond to threats more quickly, effectively, and strategically. These tools are not merely about reducing workload; they are about transforming security operations, improving resilience, and empowering security professionals to focus on higher-level strategic initiatives.

Orchestrating Defense, Automating Response, and Empowering Security Teams

Security automation tools are indispensable for modern cybersecurity operations, playing a crucial role in improving the efficiency, effectiveness, and overall resilience of security programs. These tools automate repetitive tasks, streamline complex workflows, integrate disparate security technologies, and empower security teams to respond to threats with greater speed, accuracy, and consistency. These tools represent a fundamental shift in how security operations are conducted, moving away from manual, reactive processes towards automated, proactive, and intelligence-driven approaches.

Types of Security Automation Tools: A Diverse and Evolving Landscape

The landscape of security automation tools is diverse and constantly evolving, with a range of solutions designed to address specific needs and challenges within security operations. Some of the key types of security automation tools include:

→ **Orchestration Tools:** These tools focus on automating workflows and integrating different security tools and systems, enabling them to work together seamlessly and efficiently. Orchestration tools provide a centralized platform for defining, managing, and executing automated security workflows, improving coordination and collaboration across different security functions.

→ **Automation Scripts:** Custom scripts, often written in languages like Python or PowerShell, are used to automate specific security tasks, such as data collection, log analysis, or system configuration. While powerful and flexible, custom scripts can be challenging to maintain and scale, particularly in complex environments.

→ **Configuration Management Tools:** These tools automate the configuration, deployment, and management of systems and devices, ensuring consistency, reducing errors, and improving security posture. Configuration management tools are particularly valuable for managing large and complex IT environments, ensuring that systems are configured securely and in compliance with security policies.

→ **Robotic Process Automation (RPA):** RPA tools are used to automate repetitive, rule-based tasks, such as data entry, report generation, or alert triage. RPA can free up security analysts from mundane tasks, allowing them to focus on more complex and strategic initiatives.

This diverse range of security automation tools provides organizations with a variety of options for improving their security operations, depending on their specific needs, resources, and technical expertise.

Security Orchestration, Automation, and Response (SOAR): The Convergence of Orchestration, Automation, and Response

Security Orchestration, Automation, and Response (SOAR) represents a critical category of security automation tools that combines orchestration, automation, and response capabilities into a unified platform. SOAR platforms enable organizations to automate complex security workflows, integrate different security technologies, and respond to security incidents with greater speed, efficiency, and consistency. SOAR is not just about automating individual tasks; it's about orchestrating entire security operations, automating complex response actions, and empowering security teams to make more informed and strategic decisions.

SOAR Functionality: Orchestrating Defense, Automating Response, and Empowering Security Teams

SOAR platforms provide a range of core functionalities that are essential for modern security operations:

→ **Orchestration:** SOAR platforms orchestrate workflows across a wide range of security tools and systems, including security information and event management (SIEM) systems, firewalls, intrusion detection systems (IDS), intrusion prevention systems (IPS), endpoint detection and response (EDR) solutions, threat intelligence platforms (TIPs), and vulnerability management systems. This orchestration enables different security tools to work together seamlessly, share information, and coordinate response actions.

→ **Automation:** SOAR platforms automate a wide range of security tasks, including threat investigation, incident response, vulnerability remediation, and security operations center (SOC) workflows. This automation can significantly reduce the time and effort required to perform these tasks, improve accuracy, and free up security analysts to focus on more complex and strategic initiatives.

→ **Response:** SOAR platforms automate incident response actions, enabling organizations to respond to security incidents more quickly, consistently, and effectively. Automated response actions can include isolating infected devices, blocking malicious IP addresses, containing security breaches, and initiating automated investigations. This automated response capability is crucial for minimizing the impact of security incidents and reducing the time to resolution.

These core functionalities of SOAR platforms empower security teams to move beyond reactive security measures and adopt a proactive, orchestrated, and automated approach to security operations.

History of SOAR: From Siloed Automation to Orchestrated Response

The evolution of SOAR reflects the growing complexity of security operations and the increasing need for automation and orchestration to address the challenges of modern cyber threats.

Early Days: Point Solutions and Script-Based Automation: In the early days of security automation, efforts were primarily focused on automating individual tools and tasks. Security teams often developed custom scripts to automate specific functions, such as log analysis, alert triage, or vulnerability scanning. However, these point solutions and script-based automation efforts were often siloed, difficult to integrate, and challenging to scale, particularly in complex environments with a growing number of security tools.

The Rise of Security Orchestration: As security environments became more complex and the volume of security alerts increased, the need for security orchestration became apparent. Security orchestration tools emerged to provide a centralized platform for integrating different security tools and automating workflows across those tools. These tools enabled security teams to orchestrate complex security operations, such as incident response, threat hunting, and vulnerability management, improving coordination and collaboration across different security functions.

The Convergence of SOAR: SOAR platforms emerged as a natural evolution of security orchestration, combining orchestration, automation, and response capabilities into a unified platform. SOAR platforms address the challenges of security operations by providing a comprehensive solution for automating security workflows, integrating different security technologies, and responding to security incidents with greater speed, efficiency, and consistency.

SOAR Adoption and Maturity: SOAR adoption has grown significantly in recent years as organizations increasingly recognize the benefits of security automation for improving security operations, enhancing incident response capabilities, and reducing the workload of security teams. SOAR platforms are now considered an essential component of modern security operations centers (SOCs), and the SOAR market continues to mature with the development of more advanced features and capabilities.

AI and Agentic AI's Transformative Impact on SOAR: The Future of Autonomous Security Operations Artificial intelligence (AI) and particularly Agentic AI are poised to have a transformative impact on the future of SOAR, ushering in a new era of autonomous security operations. AI is not just an add-on feature for SOAR; it's a fundamental shift in how SOAR platforms operate, learn, and adapt to the evolving threat landscape.

Enhanced Automation and Intelligent Decision-Making: AI – including machine learning (ML) and deep learning – can significantly enhance SOAR platforms by automating more complex tasks, making more intelligent decisions, and improving the accuracy and efficiency of security operations. AI algorithms can analyze vast amounts of security data, identify subtle patterns, detect anomalies, and predict potential threats, enabling SOAR platforms to automate tasks that previously required human intervention, such as complex threat investigation, alert triage, and incident prioritization.

Dynamic Orchestration and Context-Aware Response: AI can improve the orchestration capabilities of SOAR platforms by dynamically adapting workflows based on the context of a security incident, the severity of the threat, and the specific characteristics of the environment. AI can analyze threat intelligence feeds, assess the potential impact of an incident, and orchestrate the most appropriate response actions in real-time, optimizing the effectiveness of security operations and minimizing the impact of attacks. This context-aware orchestration enables SOAR platforms to move beyond pre-defined workflows and adapt quickly to the dynamic nature of cyber threats.

Autonomous Response and Self-Driving Security Operations: Agentic AI, which goes beyond traditional AI by exhibiting more autonomous, goal-directed, and adaptive behavior, has the potential to revolutionize SOAR platforms and enable a future of autonomous security operations. Agentic AI could empower SOAR platforms to respond to security incidents with minimal human intervention, analyzing the situation, developing a plan of action, executing that plan autonomously, and continuously learning and adapting based on the results. This vision of self-driving security operations could significantly reduce the burden on security teams, improve response times, and enhance the overall resilience of organizations against cyberattacks.

Challenges and Considerations: Navigating the Ethical, Technical, and Security Implications of AI in SOAR

While AI and Agentic AI offer significant potential benefits for SOAR, there are also important challenges and considerations that must be addressed to ensure the responsible and effective implementation of these technologies:

→ **Bias in AI Algorithms and the Importance of Ethical AI:** AI algorithms can be biased, which can lead to unfair, inaccurate, or discriminatory outcomes in security operations. It's crucial to address bias in AI algorithms, ensure transparency and fairness in AI decision-making, and develop and enforce ethical guidelines for the use of AI in SOAR.

→ **Explainability and Transparency of AI Decisions:** It can be challenging to understand how AI algorithms make decisions, which can make it difficult to trust, validate, and audit those decisions. It's important to develop techniques for improving the explainability and transparency of AI decisions in SOAR, enabling security professionals to understand the reasoning behind automated actions.

→ **Security of AI Systems and the Need for Robust Protection:** AI systems themselves can be vulnerable to attack, which could have significant consequences for security operations. It's essential to implement robust security measures to protect AI systems from manipulation, data poisoning, and other forms of attack, ensuring the integrity and reliability of AI-driven SOAR platforms.

These challenges and considerations highlight the importance of careful planning, ethical considerations, and robust security practices in the implementation of AI and Agentic AI in SOAR.

Musashi emphasizes the importance of adaptability, continuous learning, and embracing new technologies to stay ahead of the competition. AI and Agentic AI represent a significant and transformative shift in the cybersecurity landscape, and security professionals must adapt, learn, and embrace these technologies to remain effective in the face of evolving cyber threats. SOAR platforms, enhanced by AI and Agentic AI, will play a crucial role in enabling organizations to automate security operations, orchestrate complex workflows, respond to threats more effectively, and ultimately achieve a higher level of security resilience in the digital age.

Partial List of SOAR Tools

1. **Splunk SOAR (formerly Phantom):** A robust SOAR platform known for its automation and orchestration capabilities.[https://www.splunk.com/en_us/software/soar.html]

2. **IBM Security SOAR (Resilient):** A widely adopted SOAR solution offering strong incident response and case management features. [https://www.ibm.com/products/resilient-soar]

3. **Palo Alto Networks Cortex XSOAR:** A comprehensive SOAR platform with a strong focus on extended security orchestration and automation. [https://www.paloaltonetworks.com/cortex/xsoar]

4. **Swimlane:** A security automation platform that offers SOAR capabilities with a focus on low-code automation. [https://swimlane.com/]

5. **ServiceNow Security Operations:** While known for ITSM, ServiceNow also provides security orchestration, automation, and response capabilities. [https://www.servicenow.com/products/security-operations.html]

6. **Torq.io:** A cloud-native SOAR platform that emphasizes no-code automation and ease of use.[https://torq.io/]

7. **Siemplify (acquired by Google Cloud):** A SOAR platform focused on threat management and security operations. Now part of Google Cloud's security offerings.

8. **Rapid7 InsightConnect:** A SOAR solution that integrates with Rapid7's security suite and other third-party tools. [https://www.rapid7.com/products/insightconnect/]

The Human Element

Security Awareness Training and Culture - Building the Human Firewall: Musashi emphasizes the vital importance of discipline, unwavering focus, and unyielding mental fortitude in the face of adversity. He wisely states, "Perception is strong and sight weak." In the realm of cybersecurity, the human element stands as both the strongest and the weakest link in the defense chain. Security awareness training and the cultivation of a robust security culture are not merely important; they are absolutely essential for building a resilient human firewall – a workforce that is not only aware of potential threats but also actively contributes to the organization's security posture, capable of withstanding the ever-evolving onslaught of social engineering attacks and other cyber threats.

The Critical Importance of the Human Element - Recognizing Our Role in Cybersecurity: The human element plays a pivotal role in cybersecurity, influencing both the effectiveness of security measures and the vulnerability of an organization to attacks.

Social Engineering – Exploiting Human Psychology: Attackers frequently and effectively target the human element through sophisticated social engineering tactics. These tactics, which include phishing, pretexting, baiting, and a multitude of other manipulative techniques, are designed to exploit human psychology, tricking individuals into divulging sensitive information, clicking on malicious links, or taking actions that ultimately compromise the organization's security. Social engineering attacks often bypass technical security controls, highlighting the critical need for a well-trained and vigilant workforce.

Insider Threats – The Danger Within: Insider threats, whether malicious or unintentional, present a significant and often underestimated risk to organizations. Malicious insiders may intentionally steal or sabotage data, while unintentional insiders may cause security breaches through negligence, lack of awareness, or simple human error. Addressing insider threats requires a combination of technical controls, strong policies, and a culture of security awareness that encourages employees to report suspicious activity and adhere to security best practices.

Human Error – The Inevitable Factor: Human error is an unavoidable reality and a major contributing factor to a significant number of security incidents. Simple mistakes, such as clicking on a phishing link, using weak or reused passwords, mishandling sensitive data, or falling victim to social engineering, can have severe and far-reaching consequences for an organization. Reducing human error requires a multi-faceted approach that includes comprehensive training, user-friendly security policies, and a culture that emphasizes vigilance and accountability.

Security Awareness Training – Empowering the First Line of Defense: Security awareness training is not merely a checkbox exercise; it's a fundamental investment in an organization's security posture. It serves the crucial purpose of educating employees about potential security risks, equipping them with the knowledge and skills necessary to recognize and respond to threats effectively, and reinforcing their role as active participants in protecting the organization's valuable assets.

Purpose and Objectives: The primary purpose of security awareness training is to empower employees to become a strong first line of defense against cyber threats. This involves achieving several key objectives:

→ **Raising Awareness:** Increasing employee awareness of the current threat landscape, common attack vectors, and the potential consequences of security breaches.

→ **Recognizing Threats:** Training employees to identify and recognize social engineering tactics, phishing emails, malicious links, and other common threats.

→ **Promoting Best Practices:** Educating employees on security best practices, such as password security, data handling, and incident reporting.

→ Fostering a Security Mindset: Cultivating a security-conscious culture where employees understand their individual responsibility for protecting organizational assets.

Key Training Topics: Essential Knowledge for a Secure Workforce

Effective security awareness training programs typically cover a range of essential topics, including:

→ **Phishing Awareness – Recognizing and Avoiding Deception:** Phishing awareness training is paramount, focusing on educating employees about the various forms of phishing, how to identify phishing emails and websites, and how to avoid falling victim to these deceptive attacks. Training should cover techniques for verifying sender authenticity, recognizing suspicious links and attachments, and reporting potential phishing incidents.

→ **Password Security – The Foundation of Access Control:** Password security training emphasizes the importance of creating strong, unique passwords, avoiding password reuse, and utilizing password managers to securely store and manage credentials. Employees should be educated on the risks of weak passwords, the dangers of password sharing, and the benefits of multi-factor authentication.

→ **Data Protection – Safeguarding Sensitive Information:** Data protection training focuses on educating employees about the organization's data security policies and procedures, including how to handle sensitive information securely, both online and offline. Training should cover topics such as data classification, data encryption, secure data storage, and the proper disposal of sensitive data.

→ **Mobile Security – Securing Devices and Data on the Go:** Mobile security training addresses the unique security challenges associated with mobile devices, including the risks of unsecured Wi-Fi, mobile malware, and data loss. Employees should be trained on how to protect their mobile devices, secure their data, and avoid mobile phishing attacks.

→ **Incident Reporting – The Importance of Vigilance and Communication:** Incident reporting training emphasizes the importance of promptly reporting any suspected security incidents or suspicious activity. Employees should be educated on how to identify and report security concerns, who to contact, and the importance of timely reporting for effective incident response.

Effective Training Strategies: Engaging, Relevant, and Reinforced

To maximize the effectiveness of security awareness training, programs should incorporate several key strategies:

→ **Engaging and Interactive Methods:** Effective training goes beyond passive lectures or presentations. It incorporates engaging and interactive methods, such as gamification, simulations, quizzes, and real-world scenarios, to keep employees interested, motivated, and actively involved in the learning process.

→ **Relevant and Tailored Content:** Training should be tailored to the specific risks and challenges faced by the organization and its employees. This involves identifying the most relevant threats, customizing training content to address specific roles and responsibilities, and using real-world examples that resonate with employees' daily work experiences.

→ Reinforcement and Continuous Learning: Security awareness training is not a one-time event; it's an ongoing process. Key concepts should be reinforced through regular reminders, ongoing training sessions, phishing simulations, and other activities to maintain awareness, promote continuous learning, and ensure that security best practices become ingrained in employee behavior.

Security Culture: Fostering a Shared Responsibility for Security

Security culture is not just about training; it's about creating a shared understanding, a collective responsibility, and a set of shared values, beliefs, and behaviors within an organization regarding security. A strong security culture fosters a sense of ownership, accountability, and proactive engagement in security at all levels of the organization, from the leadership team to individual employees.

Definition and Importance: Security culture can be defined as the shared values, beliefs, attitudes, and behaviors of an organization and its employees concerning security. A strong security culture is characterized by:

→ **Shared Understanding:** Employees understand the importance of security and their role in protecting the organization's assets.

→ **Collective Responsibility:** Employees feel a sense of ownership and accountability for security, recognizing that it's everyone's responsibility.

→ **Proactive Engagement:** Employees are actively engaged in security, reporting concerns, following best practices, and contributing to a secure environment.

A strong security culture is crucial for creating a resilient security posture, as it empowers employees to be vigilant, proactive, and responsible in their daily actions, significantly reducing the risk of security incidents.

Building a Strong Security Culture: Key Strategies

Building a strong security culture requires a sustained and concerted effort involving several key strategies:

→ **Leadership Support and Commitment:** Leadership must demonstrate a strong and visible commitment to security, communicating its importance to employees, providing resources for security initiatives, and setting the tone for a security-conscious organization. Leaders should actively champion security, participate in security awareness activities, and hold themselves and others accountable for security performance.

→ **Open Communication and Feedback:** Fostering open communication channels is essential for building a strong security culture. Employees should feel comfortable reporting security concerns, providing feedback on security policies and procedures, and asking questions about security best practices. Organizations should actively solicit feedback, encourage open dialogue, and create a safe space for employees to raise security-related issues without fear of reprisal.

→ **Positive Reinforcement and Recognition:** Recognizing and rewarding employees for following security best practices, reporting security incidents, and contributing to a secure environment is crucial for reinforcing positive behaviors and fostering a security-conscious culture. Organizations should implement programs to recognize and reward security champions, celebrate security successes, and publicly acknowledge employees who go above and beyond to protect the organization.

→ Continuous Improvement and Adaptation: Building a strong security culture is an ongoing process that requires continuous improvement and adaptation. Organizations should regularly assess and evaluate their security culture, solicit feedback from employees, and make adjustments to security policies, procedures, and training programs based on the evolving threat landscape and the changing needs of the organization.

Musashi emphasizes the importance of mental discipline, preparedness, and continuous improvement in the pursuit of mastery. Security awareness training and a strong security culture instill this discipline and preparedness in employees, empowering them to be an active and vital part of the organization's defense. By fostering a security-conscious workforce, organizations can significantly reduce their risk of security incidents, enhance their overall security posture, and create a more resilient and secure environment for their employees, customers, and stakeholders.

There is a great article by Oz Alashe, CEO and Founder, Cybsafe (https://www.cybsafe.com/blog/the-dogma-of-security-awareness/?=) in which he critiques traditional approaches to security awareness training, arguing that they often rely on ineffective dogmas. It challenges the idea that simply teaching employees to recognize phishing or comply with rules leads to meaningful behavior change. The article advocates for a shift towards focusing on culture and behavior, emphasizing the need to move beyond awareness as the sole goal and instead create a security-conscious environment where positive behaviors are encouraged and reinforced.

With that in mind, the approach taken in this chapter strongly aligns with the critique of security awareness dogma presented, moving beyond simply advocating for rote training on recognizing phishing and appropriate password security. Instead, it emphasizes the importance of fostering a strong security culture, akin to Alashe's call for behavior and culture change. Both approaches recognize the limitations of traditional awareness campaigns and advocate for creating an environment where security is a shared responsibility, ingrained in the organization's ethos, and driven by proactive engagement rather than just passive knowledge.

Mounting An Effective Social Engineering Defense

Protecting the Human Element from Deception and Manipulation

In "The Book of Five Rings," Musashi repeatedly stresses the importance of understanding your opponent's tactics and employing deception effectively. He states, "All warfare is based on deception." In cybersecurity, social engineering is a potent form of deception that manipulates individuals into compromising security. Defending against social engineering demands a robust strategy built on heightened awareness, constant vigilance, a deep understanding of manipulative techniques, and the ability to proactively recognize and effectively resist these deceptive attempts. Such a defense focuses on transforming individuals from potential points of vulnerability into formidable human firewalls capable of withstanding the ever-evolving onslaught of social engineering attacks.

Understanding Social Engineering: The Art of Human Exploitation

Social engineering is not merely a technical attack; it's a sophisticated form of psychological manipulation that exploits human vulnerabilities. It's the art of deceiving and manipulating people into willingly divulging confidential information, granting access to systems, or performing actions that compromise security. Attackers who employ social engineering tactics understand human psychology and leverage it to their advantage, exploiting emotions, biases, and vulnerabilities to achieve their malicious objectives.

Definition and Techniques: A Spectrum of Deception

Social engineering encompasses a wide range of techniques – all of which are aimed at manipulating individuals to bypass security measures and achieve unauthorized access or information disclosure. Some of the most common and effective social engineering techniques include:

→ **Phishing: The Digital Lure:** Phishing is one of the most prevalent and dangerous social engineering techniques. It involves sending fraudulent emails, messages, or other digital communications designed to deceive individuals into revealing sensitive information, such as login credentials, financial details, or personal data. Phishing attacks often use spoofed sender addresses, convincing language, and a sense of urgency to trick recipients into clicking on malicious links, opening infected attachments, or providing information on fake websites. Phishing has also evolved beyond email to include smishing (SMS phishing), vishing (voice phishing), and other forms of digital communication.

→ **Pretexting: The Fabricated Reality:** Pretexting involves creating a false scenario, fabricating a believable identity, or impersonating a trusted individual to persuade individuals to divulge information, grant access, or perform actions that they would not normally take. Pretexting attacks often involve extensive research and preparation to create a convincing narrative and exploit the victim's trust, desire to be helpful, or fear of authority. Attackers may impersonate IT support, law enforcement, or other authority figures to add credibility to their deception.

→ **Baiting: The Tempting Trap:** Baiting involves offering something enticing or promising, such as a free download, a special offer, access to exclusive content, or a desirable physical item, to lure individuals into clicking on a malicious link, providing information, or performing actions that compromise security. Baiting attacks exploit human curiosity, greed, or the desire to obtain something valuable, often leading victims to bypass their better judgment and overlook potential risks. Baiting can occur both online and offline, such as leaving infected USB drives in public places.

→ **Tailgating: The Unauthorized Entry:** Tailgating is a physical social engineering technique that involves gaining unauthorized access to a restricted area by closely following someone who has legitimate access. Tailgating exploits human courtesy or the reluctance to challenge someone who appears to belong, allowing attackers to bypass physical security controls and gain entry to secure facilities. This technique relies on the assumption that people will hold the door open for others or avoid causing a scene by questioning someone's presence.

→ **Quid Pro Quo:** Quid pro quo involves offering a service or benefit in exchange for information. Attackers might pose as technical support or conduct surveys, offering assistance to gain access to credentials or sensitive data. People are often willing to provide information if they believe they are getting something in return, making this a successful manipulation tactic.

The Human Vulnerability: Exploiting Our Cognitive Biases

Social engineering is particularly effective because it directly targets human psychology, exploiting our innate tendencies, emotions, and cognitive biases. Social engineers understand that humans are often the weakest link in the security chain, and they skillfully leverage this understanding to manipulate individuals and bypass even the most robust technical security controls. Some of the key human vulnerabilities and cognitive biases that social engineers exploit include:

→ **Trust and Authority Bias:** Social engineers often impersonate trusted individuals or organizations, such as supervisors, colleagues, IT support, or authority figures, to gain the victim's trust and leverage authority bias. People are more likely to comply with requests from those they perceive as trustworthy or authoritative, making them susceptible to manipulation.

→ **Fear, Urgency, and Scarcity:** Social engineering attacks often create a sense of fear, urgency, or scarcity to pressure victims into acting quickly without thinking critically, increasing the likelihood of making mistakes. This tactic exploits our natural tendency to avoid negative consequences or miss out on opportunities, leading us to make impulsive decisions.

→ **Curiosity, Greed, and the Illusion of Reward:** Baiting attacks and other social engineering techniques exploit human curiosity, greed, or the desire to obtain something valuable, tempting victims to click on links, provide information, or take actions without considering the risks. The promise of a reward or the allure of something enticing can cloud judgment and lead to risky behavior.

→ **Helpfulness, Politeness, and Social Pressure:** Social engineers often exploit human helpfulness, politeness, or the desire to conform to social norms, making requests that are difficult to refuse or taking advantage of the victim's reluctance to challenge someone or cause a scene. We are often socialized to be helpful and polite, making us vulnerable to those who exploit these tendencies.

→ **Cognitive Overload and Distraction:** Social engineers may use tactics to create cognitive overload or distraction, making it difficult for victims to process information effectively or recognize red flags. By overwhelming individuals with information or creating a sense of chaos, attackers can exploit our limited attention spans and increase the likelihood of errors.

Defending Against Social Engineering: A Proactive and Adaptive Strategy

Defending against social engineering requires a comprehensive, proactive, and adaptive strategy that integrates multiple layers of defense, including robust awareness programs, well-defined policies, effective technology, and the cultivation of critical thinking and vigilance among all individuals. It's about creating a resilient human firewall that is not only resistant to manipulation but also capable of actively contributing to the organization's security posture.

Awareness and Training – Building a Security-Conscious Culture: Awareness and training are fundamental pillars of social engineering defense. Educating individuals about social engineering tactics, techniques, and red flags is essential for empowering them to recognize and avoid these deceptive attacks. However, effective training goes beyond simply providing information; it aims to cultivate a security-conscious culture where vigilance and critical thinking are ingrained in everyday behavior and positive behaviors are encouraged and reinforced.

Comprehensive and Continuous Education: Providing comprehensive and continuous education on the evolving landscape of social engineering, covering a wide range of tactics and techniques, including phishing, pretexting, baiting, tailgating, quid pro quo, and emerging attack vectors. Training should be tailored to different roles and responsibilities within the organization, addressing the specific threats and vulnerabilities faced by each group.

Interactive and Engaging Learning Experiences: Moving beyond traditional lectures and presentations, effective training incorporates interactive and engaging learning experiences, such as gamification, simulations, role-playing exercises, and real-world scenarios. These methods enhance understanding, promote active participation, and allow individuals to practice identifying and responding to social engineering attempts in a safe environment.

Phishing Simulations and Testing: Regularly conducting phishing simulations and testing employees' ability to identify and avoid phishing emails and other social engineering attempts. These simulations provide valuable insights into the effectiveness of training programs and identify areas where further education or reinforcement is needed. It's crucial to use these simulations as a learning tool, providing feedback and guidance to those who fall victim to the simulated attack.

Reinforcement and Ongoing Awareness Campaigns: Security awareness training is not a one-time event; it's an ongoing process. Key concepts should be reinforced through regular reminders, ongoing awareness campaigns, newsletters, posters, and other communication channels. These efforts help to maintain vigilance, promote continuous learning, and ensure that security best practices remain top-of-mind.

Policies and Procedures – Establishing a Framework for Secure Behavior

Implementing well-defined policies and procedures provides a clear framework for secure behavior, establishes accountability, and helps to mitigate the risk of social engineering attacks. These policies should be regularly reviewed and updated to address emerging threats and best practices.

Data Handling and Access Control Policies: Implementing robust policies and procedures for handling sensitive information, including data classification, access control, data encryption, and secure data disposal is a fundamental requirement. Enforcing the principle of least privilege, granting access to data and systems only to those who require it for their job responsibilities should be part of policies, as is regularly reviewing and updating access controls to ensure they remain appropriate.

Identity Verification and Authentication Protocols: Establishing strict protocols for verifying identities – both in person and online – helps prevent pretexting, impersonation attacks, and unauthorized access. Implementing strong authentication measures, such as multi-factor authentication (MFA), adds an extra layer of security and reduces the risk of account compromise, even if credentials have been obtained through social engineering.

Communication and Information Sharing Guidelines: Developing clear guidelines for communication and information sharing, emphasizing the importance of verifying requests for information, being cautious of unsolicited communications, and avoiding the sharing of sensitive information through unsecure channels are all key aspects of establishing a secure framework. Training employees on how to identify and report suspicious communications or requests for information empowers them to take action.

Incident Reporting and Response Procedures: Establishing clear procedures for reporting suspected security incidents or social engineering attempts and encouraging employees to be vigilant and proactive in identifying and reporting potential threats matters. Providing a safe and confidential mechanism for reporting concerns and ensuring that employees understand the importance of timely reporting for effective incident response provides for a more effective response. These procedures should outline the steps to take in the event of an incident, including containment, eradication, and recovery.

Technology and Tools: Enhancing Detection and Prevention

While technology alone cannot provide a complete defense against social engineering, it can play a crucial role in enhancing detection, prevention, and mitigation efforts.

Advanced Email Security and Anti-Phishing Solutions: Implementing advanced email security solutions and anti-phishing software that utilize machine learning, artificial intelligence, and other technologies to detect and block malicious emails, identify suspicious links and attachments, and provide warnings to users about potential phishing attempts provide technological support for better outcomes. These solutions should be regularly updated to address the evolving tactics of phishers.

Web Filtering and Security Awareness Integration: Utilizing web filtering technologies to block access to known malicious websites and integrate security awareness messaging into the browsing experience supports user awareness. For example, if a user attempts to visit a potentially risky website, a warning message can be displayed, reminding them of security best practices.

Endpoint Detection and Response (EDR) Solutions: Deploying EDR solutions on endpoints to detect and respond to malicious activity that may result from successful social engineering attacks can provide visibility into endpoint behavior, identify suspicious processes, and enable rapid containment and remediation.

Security Information and Event Management (SIEM) Systems: Leveraging SIEM systems to collect and analyze security logs from various sources, identify anomalies, and detect suspicious patterns of activity may indicate a social engineering attack or its aftermath. SIEM systems can help security teams to correlate events, prioritize alerts, and respond to incidents more effectively.

Critical Thinking and Vigilance – Empowering the Human Firewall: Ultimately, the most effective defense against social engineering lies in cultivating a security-conscious culture where individuals are empowered to think critically, question requests, and remain vigilant in identifying and reporting potential threats. This requires fostering a sense of shared responsibility for security and empowering every individual to be an active participant in protecting the organization.

Promoting a Culture of Skepticism and Verification: Encouraging employees to adopt a healthy skepticism and to question requests for sensitive information or access, even if the request appears to be legitimate, builds workforce awareness. Emphasizing the importance of verifying the identity of the requester, confirming the legitimacy of the request through independent channels, and being cautious of unsolicited communications or unexpected requests are all key factors.

Empowering Individuals to Challenge Authority: Creating a culture where it's acceptable and even encouraged to challenge authority or question requests that seem suspicious, without fear of reprisal is essential to training in responsible behavior. Emphasize that security is everyone's responsibility and that it's better to be cautious than to comply blindly.

Developing Critical Thinking and Security Intuition: Providing training and resources will help individuals develop their critical thinking skills, security intuition, and ability to identify red flags. This includes training on how to analyze information, evaluate context, recognize manipulation tactics, and make sound judgments in security-related situations.

Fostering a Sense of Shared Responsibility and Ownership: Cultivating a sense of shared responsibility and ownership for security provides every individual with the understanding that their role in protecting the organization's assets matters and leaves them feeling empowered to contribute to a secure environment. This involves promoting open communication, encouraging feedback, and recognizing and rewarding security-conscious behavior.

Musashi's emphasis on anticipating your opponent's moves, understanding their tactics, and using deception strategically is directly applicable to defending against social engineering. It requires a similar strategic approach: understanding the tactics used by social engineers, anticipating their attempts, implementing proactive and adaptive measures to protect against them, and empowering individuals to become a resilient and vigilant human firewall. By combining comprehensive awareness, robust policies, effective technology, and a culture of critical thinking and shared responsibility, organizations can build a strong and adaptive defense against social engineering, effectively mitigating the risks posed by these pervasive and evolving attacks.

Note: I considered not using the term "Human Firewall" since some security professionals believe that is not a thing, nor is it useful. So, I used Gemini Advance Deep Research to explore the history and viewpoints on its usage. The full Gemini report is in Appendix A: Human Firewall Cybersecurity Debate. It gave me enough information to keep the term in the book.

Staying Ahead

Industry Certifications and Continuous Learning

In the relentless ebb and flow of cyber security, our capacity to defend against those who seek to compromise our systems is inextricably linked to our knowledge, skills, and abilities. Just as Musashi emphasizes the necessity for unwavering awareness and preparedness, the cyber strategist must relentlessly pursue learning and adaptation to maintain their effectiveness in this dynamic battlefield.

Think of industry certifications not merely as credentials, but as tangible proof of your dedication to your profession. The intrinsic value derived from achieving a certification lies in the acknowledgment of being a knowledgeable and skilled professional in your chosen field. The effort demanded to study, the financial investment in study materials, and the cost of the examination itself, collectively underscore your commitment to self-improvement and professional growth.

Moreover, it's a reality that numerous companies view certifications as a fundamental requirement for employment. These certifications often serve as a tangible validation of expertise, making them more than just a personal achievement, but a strategic career asset.

However, a critical question arises: how does one navigate the myriad of certifications to choose the ones that align with their strategic objectives? Should you prioritize vendor-neutral certifications that offer a broad foundation, or technology-specific certifications that provide deep expertise in a niche area? This decision, in itself, demands strategic thinking.

The initial step in resolving these questions involves a meticulous identification of the precise skill set necessary to effectively perform the functions of your current role. If your current position entails security management, with responsibilities for an organizational program, then certifications such as the Certified Information Security Manager (CISM) from ISACA or the Certified Information Systems Security Professional (CISSP) from (ISC)^{2}® would likely be appropriate, assuming you possess the requisite experience to qualify for application.

And here's a strategic tip: certifications can be strategically leveraged as part of your professional marketing strategy. It's a common occurrence in many organizational cultures for management to place a higher value on formal educational accomplishments over practical, real-world experience and expertise. In such environments, certifications can be instrumental in establishing credibility and demonstrating competence.

Remember, certifications, while holding value, are ultimately tools – another component in the cyber-strategist's arsenal within the broader

Key of Knowledge. It is how you wield these tools that truly matters. Use them judiciously, maintain a sharp focus on continuous learning, and never cease in your pursuit of expanding your knowledge, skills, and abilities. Because in the realm of cybersecurity, the only constant is the inevitability of change, and the only path to sustained success lies in the relentless pursuit of enhancing your capabilities.

Tracking Emerging Threats and Trends

In the ever-shifting and often turbulent landscape of cybersecurity, maintaining a proactive security posture transcends the mere deployment of the latest tools or the rote adherence to compliance mandates. It necessitates an unyielding pursuit of knowledge, a profound awareness of the constantly evolving threat landscape, and the cultivated ability to anticipate and prepare for future challenges. Much like a seasoned warrior meticulously scanning the horizon for signs of approaching danger, the astute cyber strategist must cultivate a state of perpetual vigilance, remaining acutely attuned to emerging threats and nascent trends.

The cyber threat landscape exists in a state of perpetual flux, characterized by relentless change and adaptation. Threat actors, ranging from individual hackers to sophisticated nation-state groups, are engaged in a continuous process of developing and refining their tactics, techniques, and procedures (TTPs) to exploit vulnerabilities, circumvent existing defenses, and achieve their malicious objectives. To effectively counter these evolving threats, cyber strategists must cultivate a deep and comprehensive understanding of prevailing trends and, crucially, possess the foresight to anticipate and prepare for future developments. This proactive approach involves far more than simply passively consuming security blogs or attending industry conferences; it demands the implementation of a systematic and rigorous approach to threat intelligence gathering, meticulous analysis, and timely dissemination of actionable insights.

One of the most significant and persistent challenges in the field of cybersecurity is the sheer speed and agility with which threats evolve. What was considered a cutting-edge attack vector or a novel exploitation technique yesterday can rapidly become obsolete, rendered ineffective as threat actors demonstrate remarkable adaptability and swiftly refine their methods to maintain their effectiveness. This rapid pace of change and the constant state of flux necessitate an unwavering commitment to continuous learning, proactive adaptation, and fostering a culture of innovation and experimentation within security teams. Cyber strategists must actively and aggressively seek out new information, experiment with novel and unconventional defense strategies, and cultivate an organizational environment that encourages innovation and proactive problem-solving.

Emerging threats and trends can manifest themselves in a multitude of forms, each presenting unique challenges and requiring tailored defense strategies. These threats and trends may include, but are not limited to:

→ **The Relentless Rise of Ransomware and Malware:** Ransomware attacks continue to pose a significant and escalating threat to organizations of all sizes and across all sectors. Threat actors are not only persisting with ransomware attacks but are also constantly developing new and more sophisticated ransomware variants with enhanced capabilities, such as advanced encryption algorithms, data exfiltration capabilities for extortion, and the use of "double extortion" tactics to increase pressure on victims to pay ransoms. Similarly, malware – in its various forms – remains a persistent and pervasive threat, with attackers employing increasingly sophisticated and evasive techniques to bypass traditional detection mechanisms and maintain persistence within compromised   systems.

→ **Escalating Cloud-Based Threats and Vulnerabilities:** The widespread and accelerating adoption of cloud computing technologies and services has introduced a new set of complex and multifaceted security challenges. Cloud-based threats, such as large-scale data breaches resulting from misconfigurations, unauthorized access due to compromised credentials, denial-of-service attacks targeting cloud infrastructure, and data leakage due to inadequate security controls, require specialized security measures, cloud-specific security expertise, and a deep understanding of cloud service provider security models and responsibilities.

→ **The Persistent Challenge of Social Engineering Tactics and Defense:** Social engineering, the art of manipulating human psychology to deceive individuals into divulging sensitive information or performing actions that compromise security,[1] remains a highly effective and frequently employed attack vector. As social engineering tactics become increasingly sophisticated, personalized, and persuasive, organizations must invest heavily in robust and comprehensive security awareness training programs that educate employees about the latest social engineering techniques and implement effective defense mechanisms, including strong authentication protocols, strict access controls, and a culture of security awareness and skepticism.

→ **The Enduring Threat of Insider Threats and Mitigation:** Insider threats, which can originate from malicious intent or unintentional negligence, continue to pose a significant and often underestimated risk to organizations. Detecting and effectively mitigating insider threats requires a multi-layered approach that combines technical controls, such as robust access management systems, data loss prevention (DLP) technologies, and user activity monitoring, with organizational measures, such as thorough background checks, stringent security policies and procedures, and comprehensive security awareness training that emphasizes the importance of protecting sensitive information and reporting suspicious activity.

→ **The Ubiquity of Phishing and the Importance of Email Security:** Phishing attacks, delivered through various channels but predominantly through email, remain a primary and highly effective method for attackers to gain initial access to systems, networks, and sensitive data. Organizations must implement robust and multi-faceted email security measures, including advanced spam filters, sophisticated anti-phishing technologies that leverage

AI and machine learning, and comprehensive employee training programs that educate users about how to identify and avoid phishing attempts, to effectively defend against these pervasive and evolving   attacks.

→ **The Growing Complexity of Web Application Attacks and Security:** Web applications have become a ubiquitous and critical component of modern business operations, but they also represent a significant and attractive target for attackers. Web application attacks, such as SQL injection, cross-site scripting (XSS), broken authentication mechanisms, and API vulnerabilities, are on the rise, and securing web applications requires a combination of secure development practices implemented throughout the software development lifecycle (SDLC), regular and thorough vulnerability scanning and penetration testing, and the deployment of web application firewalls (WAFs) to detect and block malicious traffic.

→ **The Ever-Present Risk of Data Breaches and the Necessity of Data Loss Prevention:** Data breaches, which can result from a variety of attack vectors, including hacking, insider threats, and social engineering, can have devastating consequences for organizations, including significant financial losses, severe reputational damage, legal liabilities, and regulatory penalties. Organizations must implement robust and comprehensive data loss prevention (DLP) strategies and technologies to protect sensitive data from unauthorized access, use, or disclosure, both internally and externally, and to ensure compliance with relevant data privacy regulations.

→ **Emerging Threats and the Challenges of the Future (AI-powered attacks, quantum computing):** The rapid emergence and advancement of new and disruptive technologies, such as artificial intelligence (AI) and quantum computing, present both unprecedented opportunities and significant challenges for the field of cybersecurity. AI can be leveraged to enhance threat detection, automate security operations, and improve incident response, but it can also be exploited by attackers to develop more sophisticated, automated, and evasive attacks. Quantum computing, while still in its early stages, has the potential to break current encryption methods that underpin much of modern cybersecurity, requiring organizations to proactively prepare for a future post-quantum cryptography era by investing in research and development of quantum-resistant cryptographic algorithms and security protocols.

To effectively track and respond to emerging threats and trends, cyber strategists can leverage a variety of valuable resources and strategies, including:

→ **Threat intelligence platforms and services:** We talked about these earlier. These specialized platforms and services provide valuable and actionable insights into current and emerging threats, including detailed information on attacker TTPs, malware trends, vulnerability disclosures, and indicators of compromise (IOCs). Threat intelligence platforms enable security teams to proactively identify and respond to threats, prioritize security efforts, and make more informed security decisions.

→ **Industry reports and publications from reputable sources:** Security vendors, research organizations, government agencies, and industry consortia regularly publish in-depth reports and analyses on cybersecurity trends, emerging threats, and best practices. These publications provide valuable information and insights that can help cyber strategists stay informed and adapt their security strategies accordingly.

→ **Active participation in security conferences and workshops:** Attending and actively participating in security conferences, workshops, and training events provides valuable opportunities for cyber strategists to learn from industry experts, network with peers, share knowledge and experiences, and stay up-to-date on the latest research, developments, and best practices in the field.

→ **Engagement in online communities and forums:** Online communities, forums, and discussion groups dedicated to cybersecurity can be valuable resources for cyber strategists to share information, discuss security challenges, collaborate on solutions, and learn about emerging threats, vulnerabilities, and effective defense strategies from a diverse range of security professionals.

By actively and consistently tracking emerging threats and trends, and by proactively adapting their security strategies and defenses, cyber strategists can significantly enhance their organization's overall security posture, effectively anticipate and mitigate future challenges, and mount a more robust and resilient defense against the ever-evolving and increasingly sophisticated cyber adversary.

Key 1 : Summary

Key 1, Self-Knowledge, is foundational to cybersecurity strategy. Just as "The Ground Book" provides an overall picture of the art of fighting, this key emphasizes understanding the cybersecurity landscape and one's place within it.

Here's how the concepts align:

→ **Understanding Your Environment:** This involves asset classification, vulnerability management, and risk management. Just as a warrior must know the battlefield, a cyber strategist must know   their   digital   terrain.

→ **Tools of the Trade:** This includes SIEM systems, penetration testing tools, threat intelligence platforms, and security automation. Like a warrior mastering their weapons, a strategist must   master   their   tools.

→ **The Human Element:** This covers security awareness training, social engineering defense, and ethics. A warrior must understand not only their own capabilities but also the behavior of their opponent; similarly, a cyber strategist must understand the human element   in   security.

→ **Staying Ahead:** This involves industry certifications, continuous learning, and tracking emerging threats. Just as a warrior must constantly train and adapt, a cyber strategist must continuously learn and evolve.

In essence, Key 1 embodies the principles of "The Ground Book" by emphasizing the importance of a solid foundation of knowledge, awareness, and understanding. This foundation is crucial for effective cybersecurity strategy.

Key 2: Agility (The Water Book)

"You should not have a favorite weapon. To become over-familiar with one weapon is as much a fault as not knowing it sufficiently well." — Miyamoto Musashi

Musashi teaches us that water adapts to its container, whether it's a small trickle or a vast ocean. His lessons extend beyond swordsmanship, emphasizing the importance of balance, a calm demeanor, and the ability to be agile in any situation. This concept of agility is central to an effective cybersecurity strategy.

In the cyber world, agility directly influences how we defend our systems and respond to security incidents. While the subsequent "Fire Book" will delve into proactive engagement and offensive strategies, this second key, Agility, centers on our defensive capabilities and readiness. It's about cultivating the "Spiritual Bearing," adopting the right "Stance," and maintaining the appropriate "Gaze" to ensure a swift and effective response when an attack occurs.

Too often, organizations approach incident response with a reactive mindset, treating it as mere "incident reaction." In these scenarios, the primary reactions often involve a flurry of questions: "Who else knows about this?" and "Do we have to report this to anyone?" Subsequently, the focus shifts to restoring affected systems; however, this is frequently done reactively without a comprehensive understanding of the incident's scope, the extent of the damage, or – critically – whether the attacker or threat has been fully eradicated.

This reactive approach demonstrates a fundamental lack of strategic depth. True agility in cybersecurity involves developing a well-defined process firmly grounded in sound principles. It's about building a framework that allows for flexibility, adaptability, and a capacity for effective and decisive action. It's about emulating water: fluid, adaptable, and possessing the capacity to wield immense power when necessary.

Incident Response

Developing a Response Plan

NOTE: *When I was writing this section, the phrase "Not if, but when" again crossed my mind. I was on a panel at The Seattle/Bellevue Cybersecurity Summit 2025, and the topic was incident response. A fellow panelist used the phrase when describing the development of an incident response plan. I added that I don't subscribe to that fatalistic mentality, and in response, I attributed the original quote to former FBI Director James Comey. I was incorrect. While researching the history of the quote, I discovered that the FBI Director Robert Mueller was the original. You can read the full results of the research in Appendix B.*

Agility in cybersecurity transcends mere reaction to the inevitable. A modern cyber strategist operates on the principle that minimizing the frequency and impact of incidents hinges on a carefully orchestrated combination of proactive resilience and highly adaptable response capabilities. Much like water, which possesses the inherent ability to both withstand immense pressure and flow effectively when necessary, our cybersecurity strategy must embody this dual nature. Developing a comprehensive and meticulously crafted incident response plan is not simply a best practice or a compliance checkbox; it's a strategic imperative that works in close synergy with proactive security measures and resilient system design.

A well-structured incident response plan provides a clear, concise, and actionable roadmap for navigating the multifaceted complexities of a cyber event. It serves as a guide, outlining the precise roles and responsibilities of key personnel within the organization, establishing clear and efficient communication protocols to ensure information flows smoothly during a crisis, and defining the specific, sequential steps to be taken to effectively contain, thoroughly eradicate, and efficiently recover from an incident when it inevitably occurs – despite our most diligent and robust efforts at prevention. This plan acknowledges that even the most sophisticated and proactive security measures may not be foolproof, and therefore, a well-defined response mechanism is critical.

The development of a robust and effective incident response plan should commence with a comprehensive and detailed assessment of the organization's critical assets, potential threats that could target those assets, and any existing vulnerabilities that could be exploited.

This initial assessment is not merely a checklist exercise; it's a foundational step that informs the prioritization of response efforts, ensuring that the organization's most valuable and sensitive assets receive the highest level of protection through resilient system design and proactive threat mitigation strategies.

Key components of a well-defined and agile incident response plan include:

→ **Preparation:** This crucial phase involves establishing a clear and well-defined organizational structure for incident response, explicitly defining the roles and responsibilities of each team member and stakeholder and developing robust and redundant communication channels to ensure seamless information flow during an incident. It also includes conducting thorough and regular risk assessments to identify potential vulnerabilities, implementing proactive and effective security controls to minimize the likelihood of incidents, and providing comprehensive and ongoing security awareness training to all employees, empowering them to be a first line of defense. Crucially, preparation extends beyond these traditional measures to encompass the design and implementation of resilient systems that can withstand attacks and minimize damage, as well as the proactive gathering and analysis of threat intelligence to anticipate and prevent incidents before they occur, shifting the focus from purely reactive to a more proactive security   posture.

→ **Detection and Analysis:** This phase focuses on the rapid and accurate identification and thorough analysis of security incidents.

It involves continuous and vigilant monitoring of systems for any signs of suspicious activity, prompt and efficient investigation of potential incidents to determine their validity, and a comprehensive analysis to determine the precise scope and potential impact of the attack on the organization. Effective detection and analysis are significantly enhanced by proactive threat hunting activities, where security teams actively search for signs of malicious activity, and continuous monitoring for anomalies and deviations from normal behavior that may indicate an ongoing breach or an impending attack, allowing for early intervention and mitigation.

→ **Containment, Eradication, and Recovery:** This phase involves taking decisive and coordinated action to contain the incident and prevent further damage, thoroughly eradicate the threat from the affected systems and environment, and efficiently recover affected systems and data to restore normal operations. It includes isolating affected systems to prevent lateral movement, meticulously removing malware and other malicious artifacts, restoring compromised data from secure and reliable backups, and implementing robust and effective measures to prevent similar incidents from occurring in the future. Recovery efforts should not only focus on restoring systems to their previous state but also, and perhaps more importantly, on enhancing the overall resilience of those systems and the organization's security posture to better withstand future attacks and minimize their potential impact.

→ **Post-Incident Activity:** This phase focuses on meticulous documentation of the incident, a thorough and objective analysis of lessons learned from the incident response, and the implementation of concrete improvements to the incident response plan and security protocols based on those lessons. It also includes clear and timely communication with relevant stakeholders, such as customers, regulators, and law enforcement agencies, as required. Post-incident analysis should provide valuable insights that directly feed back into proactive threat intelligence gathering and resilience strategies, creating a continuous feedback loop that enables the organization to continuously improve its security posture and adapt to the evolving threat landscape.

Just as water adapts to the shape of its container and flows around obstacles, the incident response plan must be inherently adaptable and flexible to accommodate the diverse range of potential incidents and the ever-changing nature of cyber threats. It should be regularly and rigorously reviewed, tested, and updated to accurately reflect changes in the threat landscape, advancements in technology, and evolving business operations. However, this agility and adaptability are most effective and impactful when paired with resilient system design and architecture that minimizes the potential impact of incidents in the first place, reducing the reliance on purely reactive measures.

In addition to having a well-defined and documented plan, it is absolutely critical to practice its execution regularly. Tabletop exercises, realistic simulations, and comprehensive drills can help to identify potential weaknesses and gaps in the plan, and ensure that all personnel involved are adequately prepared to respond effectively and efficiently during a real-world incident. These exercises should also be designed to rigorously test the resilience of systems and the effectiveness of proactive security measures, providing a holistic evaluation of the organization's security readiness.

Developing and maintaining an effective incident response plan is not a one-time activity or a static document; it's an ongoing process that requires continuous improvement, constant adaptation, and an unwavering commitment to learning from past incidents, both internal and external. By embracing agility, prioritizing the design and implementation of resilient systems, and proactively seeking out and mitigating potential threats, cyber strategists can effectively navigate the turbulent and unpredictable waters of cybersecurity incidents and protect their organizations from significant harm.

The OODA Loop in Action: Embodying the Essence of the Water Book

What follows is a short discussion of OODA. I have written more extensively about it in my book Unleashing the Power of the OODA Loop in Cybersecurity: A New Frontier in Cybersecurity: Applying the OODA Loop and Implicit Guidance and Control.

The OODA Loop, conceived initially by military strategist John Boyd, provides an invaluable framework for understanding and significantly improving incident response agility within cybersecurity. The player who navigates these steps most rapidly and efficiently, demonstrating superior adaptability, often emerges as the victor in any engagement. This principle holds true whether on the battlefield or in the digital realm.

In the context of "The Water Book" and its emphasis on agility, the OODA Loop exemplifies the fluid and adaptable nature required to effectively respond to cybersecurity incidents. Just as water adapts to its container, changing shape and direction as needed, cybersecurity professionals must leverage the OODA Loop to navigate the dynamic and often unpredictable landscape of cyberattacks.

Here's a high-level breakdown of how the OODA Loop translates to cybersecurity incident response:

Observe: This initial stage involves continuously and vigilantly monitoring the cybersecurity environment for potential threats, vulnerabilities that could be exploited, and any indicators of compromise that may signal an ongoing attack. In cybersecurity, this translates to leveraging a wide array of sophisticated tools and technologies, including Security Information and Event Management (SIEM) systems, Intrusion Detection Systems (IDS), comprehensive threat intelligence feeds, and various other monitoring and analysis tools that have been introduced in previous sections. These tools enable security teams to gather a continuous stream of data regarding network activity, detailed system logs, application behavior, and potential security events that may warrant further investigation. This phase is about maintaining a state of heightened awareness, much like a warrior sensing the subtle shifts in their   opponent's   stance.

Orient: This critical step transforms the raw data gathered during the "Observe" phase into actionable intelligence. It involves a deep and thorough analysis of the information, placing it within a relevant context, and determining its potential impact on the organization's systems and data. This complex process requires skilled security analysts to correlate seemingly disparate events, identify patterns that may indicate malicious activity, and prioritize alerts based on their severity and potential impact. The goal of this step is to develop a comprehensive understanding of a security incident's nature, scope, and possible consequences, enabling informed decision-making in the subsequent phases. This is akin to a warrior not just seeing their opponent but truly understanding their intent and strategy.

Decide: Based on the insights gained during the "Orient" phase, security teams must make critical decisions regarding the most appropriate course of action to effectively respond to the incident. This involves developing a comprehensive and strategic response strategy, carefully selecting the most suitable tools and techniques to mitigate the threat and efficiently assigning tasks to the appropriate personnel based on their expertise and responsibilities. This decision-making process requires a clear understanding of the organization's incident response plan and the ability to adapt and make real-time adjustments based on the evolving situation.

Act: This is the phase where the carefully crafted response plan is put into action. It involves taking decisive and coordinated steps to contain the incident and prevent further damage, thoroughly eradicate the threat from the affected systems and environment and efficiently recover compromised systems and data to restore normal operations. This phase demands swiftness, precision, and effective coordination among various teams and stakeholders to minimize the impact of the incident and ensure a timely and complete recovery. This is the equivalent of a warrior executing a precise and decisive strike, aiming to neutralize the threat as quickly and efficiently as possible.

The OODA Loop is not a static, linear process with a defined beginning and end; it's a dynamic and continuous cycle emphasizing iteration and learning. The results and outcomes of the "Act" phase provide valuable feedback that flows back into the "Observe" phase, enabling security teams to continuously learn from each incident, refine their understanding of threats, and proactively improve their overall incident response capabilities. This iterative process allows organizations to become more agile and adaptive in their security posture, better prepared to face future challenges.

Agile Security Frameworks and DevSecOps: Weaving Security into the Development DNA, the "Water" Way

The traditional approach to security, often characterized by a rigid and siloed "throw it over the wall" mentality, simply doesn't hold water (pun intended) in today's dynamic and fast-paced development environment. Agile security frameworks and DevSecOps represent a fundamental and essential shift in integrating security into the software development lifecycle (SDLC). It's a move away from treating security as an afterthought and towards recognizing it as an integral and inseparable aspect of the entire development process.

Agile methodologies, born from the pressing need for rapid adaptation and flexibility in software development, emphasize iterative development cycles, close collaboration between teams, and the importance of continuous feedback. When these principles are applied to security, agile frameworks promote:

Flexibility and Adaptability: Security practices must possess the inherent ability to adapt and evolve in response to the ever-changing needs of the development process and the constantly shifting threat landscape. Like water flowing around obstacles, security must be fluid and responsive.

Rapid and Iterative Response: Security teams need to provide quick and actionable feedback to development teams and be prepared to adapt security measures in real-time, keeping pace with the speed of development.

Continuous Improvement and Learning: Security should not be viewed as a static set of rules but rather as a dynamic and evolving process that requires continuous evaluation, refinement, and improvement throughout the entire development lifecycle.

Collaboration and Shared Responsibility: Breaking down traditional silos between development, operations, and security teams is essential to foster a culture of shared responsibility for security, where everyone understands their role in maintaining a strong security posture.

DevSecOps takes these principles a significant step further by fully integrating security into every single stage of the DevOps pipeline. DevOps, a set of practices that automates the processes between software development and IT teams, enables faster and more reliable software releases. DevSecOps builds upon this foundation by "shifting left," meaning that security considerations are proactively addressed from the very beginning of the development process, rather than being reactively bolted on as an afterthought at the end.

To illustrate this concept, think of it this way: instead of constructing a house and then hiring a security specialist to install security cameras, alarms, and reinforce entry points, DevSecOps is akin to designing the house with security as a primary consideration from the initial blueprint, incorporating reinforced walls, secure entry points, and a comprehensive, built-in surveillance system.

The key benefits of embracing Agile security frameworks and DevSecOps include:

Faster Time to Market: By integrating security early and throughout the development process, vulnerabilities are identified and addressed more rapidly and efficiently, significantly reducing delays in the software release cycle and enabling quicker delivery of secure products.

Improved Overall Security Posture: Security becomes a shared responsibility that is deeply ingrained in the development process, leading to the development of inherently more secure code and robust systems.

Increased Efficiency and Streamlining: Automation and enhanced collaboration streamline security processes, freeing up valuable resources and allowing teams to focus on other critical priorities, optimizing the use of time and effort.

Enhanced Communication and Collaboration: DevSecOps fosters improved communication, stronger collaboration, and tighter integration between development, operations, and security teams, resulting in a more cohesive, coordinated, and ultimately more effective approach to security.

However, it's important to acknowledge that adopting Agile security frameworks and DevSecOps is not without its inherent challenges. It requires a significant and often complex cultural shift within the organization, necessitating the breaking down of traditional silos and the cultivation of a strong sense of shared responsibility for security across all teams. It also necessitates a strategic investment in new tools, automation technologies, and comprehensive training programs to equip teams with the necessary skills and knowledge to effectively implement and succeed with these new approaches.

Ultimately, Agile security frameworks and DevSecOps are not merely about implementing a set of new tools or processes. They represent a fundamental and profound change in mindset, a recognition that security is not an obstacle or a roadblock to innovation and speed, but rather a critical enabler that allows organizations to achieve both speed and security simultaneously. By embracing agility, fostering adaptability, and seamlessly integrating security into the software development lifecycle, organizations can thrive in today's fast-paced digital world, building secure and resilient systems that meet the demands of both the business and the ever-evolving threat landscape.

Cloud Security and Adapting to New Technologies: Embracing a Shared, Fluid, and Strategic Approach

The rapid and widespread shift to cloud computing has ushered in a profound paradigm shift in how we approach and implement security. It's no longer sufficient to focus solely on protecting a static and well-defined perimeter; instead, it demands a fundamentally different approach centered on securing a highly dynamic, distributed, and often complex environment. This necessitates a mindset that wholeheartedly embraces adaptability, mirroring the way water effortlessly takes the shape of its container, along with a clear and comprehensive understanding of the shared responsibility model that underpins cloud security.

The Shared Responsibility Model: A Collaborative Foundation

In the cloud computing paradigm, security is not a singular responsibility but rather a collaborative effort shared between the cloud provider and the cloud user. This shared model is essential for establishing a robust and effective security posture.

The Cloud Provider's Responsibility: The cloud provider assumes responsibility for the security of the cloud infrastructure itself. This encompasses a wide range of critical security functions, including the physical security of the data centers where the cloud infrastructure resides, the security of the underlying hardware and software infrastructure that powers the cloud, and the security of the cloud platform and its core services.

The Cloud User's Responsibility: Conversely, the cloud user is responsible for security in the cloud environment. This crucial responsibility includes securing their own data that is stored and processed within the cloud, securing the applications they deploy and run in the cloud, and diligently managing the configurations of their cloud resources and services to ensure they adhere to security best practices.

A clear understanding and acceptance of this shared responsibility model is absolutely critical. It fundamentally means that organizations cannot afford to simply assume that the cloud provider will handle all aspects of security automatically. They must proactively and actively engage in securing their own assets, managing their own applications, and diligently configuring their cloud environments to maintain an acceptable level of security.

Adapting to New Technologies: Embracing Fluidity and Continuous Learning

The cloud environment is characterized by constant evolution, with new technologies, services, and deployment models emerging at an incredibly rapid pace. To maintain a strong and effective security posture, organizations must cultivate the ability to adapt quickly and effectively to these changes. This requires a commitment to:

→ **Agility and Adaptability:** Security practices, controls, and strategies must be inherently flexible, adaptable, and capable of evolving in response to the ever-changing landscape of cloud technologies and deployment models. Much like water adapts to the shape of its container, security must be fluid, responsive, and able to adjust to new circumstances.

→ **Continuous Learning and Skill Development:** Security professionals must commit to continuous learning, staying up-to-date on the latest cloud security best practices, emerging threats, and new cloud technologies. This requires a proactive approach to professional development and a willingness to acquire new skills.

→ **Automation and Orchestration:** Automating security tasks and processes is essential for keeping pace with the speed, scale, and complexity of the cloud environment. Automation can help to improve efficiency, reduce errors, and enhance the overall effectiveness of security operations.

Cloud security and adapting to new technologies also necessitate embracing a "Water Book" approach by drawing inspiration from its principles of fluidity, balance, and strategic thinking:

→ **Fluidity and Dynamic Adaptation:** Security controls and strategies should be fluid, dynamic, and adaptable, capable of changing and evolving as the cloud environment changes, new threats emerge, and new technologies are adopted. This requires a flexible   and   proactive   security   mindset.

→ **Balance Between Security and Business Needs:** Maintaining a delicate balance between security and business needs is absolutely crucial. Security should be viewed as an enabler of innovation and business growth, not as an inhibitor or a roadblock. A strategic approach is needed to integrate security seamlessly into business processes.

→ **Strategic Thinking and Proactive Planning**: A strategic and proactive approach is essential for effectively navigating the complexities of cloud security and adapting to new technologies. This involves anticipating future challenges, developing long-term security plans, and making informed decisions about security investments.

Some of the key challenges and critical considerations that organizations must address in cloud security include:

→ **Data Security and Privacy**: Protecting sensitive data stored and processed in the cloud is of paramount importance. Organizations must carefully and strategically consider data encryption methods, implement robust access control mechanisms, and employ effective data loss prevention (DLP) strategies to safeguard their valuable information.

→ **Identity and Access Management (IAM):** Implementing robust and comprehensive IAM practices is crucial for controlling access to cloud resources and preventing unauthorized access. This involves implementing strong authentication, authorization, and access control policies and procedures.

→ **Compliance with Regulations and Standards:** Organizations must ensure that their cloud deployments and security practices comply with all relevant legal, regulatory, and industry-specific compliance requirements, such as HIPAA, PCI DSS, GDPR, and other applicable standards.

→ **Visibility, Monitoring, and Control**: Gaining adequate visibility into and maintaining effective control over cloud environments can be a significant challenge. Organizations need to invest in appropriate tools, implement robust monitoring processes, and establish clear governance policies to detect threats, monitor activity, and enforce security policies effectively.

Navigating the complexities of cloud security effectively and successfully adapting to new technologies requires a proactive, agile, and strategic approach. It's about fully embracing the shared responsibility model, committing to continuous learning and improvement, and implementing security measures as dynamic, adaptable, and resilient as the cloud.

Adapting to AI: Navigating a Dual-Use Revolution with Agentic Implications

Artificial Intelligence (AI) and its subset Machine Learning (ML) are not just incremental technological advancements; they represent a fundamental and disruptive shift that is reshaping every facet of our lives, and the realm of cybersecurity is certainly no exception. This transformation presents us with a dual and significant challenge: we must proactively adapt to both the immense benefits that AI offers for enhancing and strengthening our security defenses and the very real and potent risks it poses as a powerful and increasingly sophisticated tool in the hands of malicious actors.

AI for Cybersecurity: A Powerful Force Multiplier, But Not a Silver Bullet

AI offers tremendous and transformative potential to enhance and revolutionize cybersecurity in a multitude of ways:

→ **Enhanced Threat Detection and Analysis:** AI algorithms possess the capability to analyze massive and complex datasets with unparalleled speed, scale, and accuracy, enabling them to detect subtle patterns, anomalies, and indicators of malicious activity that would be virtually impossible for human analysts to identify in a timely manner. This capability facilitates proactive threat hunting, enabling security teams to identify and neutralize threats before they can cause significant damage, and it also allows for faster and more accurate identification of ongoing malicious activity.

→ **Automation of Security Tasks:** AI can automate a wide range of repetitive, time-consuming, and often mundane security tasks, freeing up valuable time and resources for security professionals to focus on more strategic, complex, and critical issues that require human expertise and judgment. This automation can be applied to tasks such as vulnerability scanning, incident triage, initial incident response actions, and continuous security monitoring, improving efficiency and reducing the burden on security teams.

→ **Predictive Security and Risk Assessment:** ML models can be trained on vast amounts of historical data to identify trends, predict future attacks, and anticipate potential vulnerabilities before they can be exploited by attackers. This predictive capability allows organizations to take proactive measures to strengthen their defenses, mitigate potential risks, and prevent attacks before they occur.

→ **Enhanced and Accelerated Incident Response:** AI can play a crucial role in automating and orchestrating various aspects of incident response, enabling faster, more coordinated, and more effective containment, mitigation, and recovery from attacks. AI-driven systems can analyze incident data in real-time, identify the scope and impact of an attack, and recommend or even automatically execute specific response actions, minimizing damage and downtime.

The Emergence of Agentic AI and its Implications

It's also crucial to consider the emergence of Agentic AI, a type of AI that can operate autonomously, make decisions, and take actions without explicit human direction. This presents both opportunities and risks for cybersecurity.

On the defensive side, Agentic AI could lead to highly adaptive and self-defending systems that can respond to threats in real-time with minimal human intervention.

On the offensive side, Agentic AI could enable attackers to create highly sophisticated and autonomous attack tools that can learn, adapt, and evade defenses with unprecedented effectiveness.

AI Against Cybersecurity: A Double-Edged Sword

It's critical to acknowledge that the same powerful capabilities that make AI a valuable tool for enhancing cybersecurity also make it an incredibly potent and dangerous weapon in the hands of malicious actors. This creates a challenging dynamic where AI is used both to defend and to attack, leading to a continuous escalation in the sophistication of cyber warfare.

AI-Powered and Enhanced Attacks: Attackers can leverage AI to automate and scale their attacks, making them more sophisticated, more targeted, and significantly more difficult to detect and defend against. This includes the development of AI-driven phishing campaigns that are highly personalized and convincing, the creation of malware that can evade traditional detection methods by adapting its behavior, and the automation of vulnerability exploitation to rapidly identify and compromise vulnerable systems.

Evasion of Security Defenses: AI can be employed to develop attacks that are specifically designed to evade and bypass existing security defenses, such as intrusion detection systems, firewalls, and endpoint protection solutions. Attackers can use AI to analyze the behavior of these defenses and craft attacks that can effectively circumvent them.

Amplification of Attack Effectiveness: AI can amplify the effectiveness and impact of cyberattacks, allowing attackers to achieve greater damage and disruption with less effort and fewer resources. For example, AI can be used to optimize denial-of-service attacks, making them more powerful and resilient, or to automate the spread of malware, increasing the speed and scope of infection.

A Realistic Perspective: Maintaining Pace, Not Gaining an Advantage

It's my considered opinion that, in the long run, AI will not necessarily provide cybersecurity professionals with a decisive and lasting advantage over bad actors. Instead, AI will likely become a tool that both sides utilize, leading to a continuous arms race where each side develops and deploys increasingly sophisticated AI-powered tools and techniques.

Therefore, our goal should be to leverage AI effectively to maintain pace with the evolving threat landscape and avoid falling behind, rather than expecting it to provide a guaranteed and permanent upper hand.

Strategic Considerations for Adapting to AI

Adapting effectively to the challenges and opportunities presented by AI in cybersecurity requires a strategic, proactive, and nuanced approach:

→ **Embrace and Integrate AI for Defense:** Organizations must prioritize the investment in and adoption of AI-powered security tools, technologies, and strategies to enhance and strengthen their defensive capabilities. This includes the implementation of AI-driven threat detection systems, the development of automated incident response systems, and the utilization of AI-powered predictive security analytics to anticipate and prevent attacks.

→ **Proactive Preparation for AI-Powered Attacks:** Organizations must not only focus on utilizing AI for defense but also proactively anticipate and prepare for the emergence of new and sophisticated attack vectors that AI will enable. This requires the development of novel and adaptive detection and response strategies, as well as a significant investment in research and development to stay ahead of the curve in this rapidly evolving landscape.

→ **Focus on Talent Development and Skills Enhancement:** Successfully adapting to AI in cybersecurity necessitates a workforce that possesses the specialized skills and knowledge required to both develop and effectively utilize AI-powered security tools and to defend against increasingly sophisticated AI-powered attacks. Organizations must prioritize investing in comprehensive training and education programs to equip their security teams with these essential skills.

→ **Foster Collaboration and Information Sharing:** Robust collaboration and effective information sharing are absolutely crucial for staying ahead of the curve in the face of AI-powered threats. Organizations must actively collaborate with industry partners, government agencies, and research institutions to share threat intelligence, exchange best practices, and collectively develop effective countermeasures.

→ **Address Ethical Considerations and Responsible Use:** The increasing use of AI in cybersecurity raises important and complex ethical considerations, including concerns related to privacy, potential bias in AI algorithms, and the critical need for accountability in AI-driven security decisions. Organizations must carefully consider these ethical implications and proactively develop comprehensive policies and guidelines to ensure the responsible, ethical, and transparent use of AI in their security practices.

Adapting to AI in cybersecurity is not merely an option; it's an absolute necessity for ensuring organizational resilience and survival in the face of a rapidly evolving and increasingly complex threat landscape. By strategically embracing AI for defensive purposes, proactively preparing for AI-powered attacks, and adopting a comprehensive and strategic approach, organizations can harness the immense power of AI to enhance their security posture, mitigate potential risks, and strive to maintain pace in the ongoing cyber arms race.

Threat Modeling: Embracing Fluidity, Proactive Defense, and Strategic Foresight

Threat modeling is not merely a technical exercise or a compliance checkbox; it's a fundamental and indispensable aspect of a proactive cybersecurity strategy. It's a structured process for systematically identifying potential security threats, vulnerabilities, and attack vectors that could compromise a system, application, or network. This process involves a detailed analysis of the system's architecture, design, implementation, and operational environment to gain a comprehensive understanding of how it could be exploited by malicious actors. By adopting this proactive approach, security professionals can anticipate potential attacks, prioritize risks based on their likelihood and potential impact, and implement appropriate and effective security measures to mitigate those risks before they can be exploited.

The Profound Importance of Threat Modeling

Threat modeling provides a multitude of critical benefits that contribute significantly to a strong security posture:

→ **Proactive and Preventative Security:** One of the most significant advantages of threat modeling is its proactive nature. By identifying potential threats and vulnerabilities early in the development lifecycle of a system or application, threat modeling enables organizations to integrate security considerations into the design and development process from the very beginning. This "shift left" approach allows for building security into the system's foundation, rather than attempting to bolt it on as an afterthought, which is often less effective and more costly in the long run.

→ **Strategic Risk Prioritization:** Threat modeling plays a crucial role in helping organizations prioritize their security efforts and allocate resources effectively. By systematically identifying and analyzing potential threats and vulnerabilities, threat modeling allows security teams to focus their attention and resources on the most critical risks, those that pose the greatest potential harm to the organization's assets and operations. This risk-based approach ensures that security efforts are aligned with the organization's priorities and that resources are used efficiently.

→ **Enhanced Security Design and Architecture:** The insights and understanding gained from the threat modeling process can be invaluable in improving the security design and architecture of systems and applications. By identifying potential weaknesses and vulnerabilities, threat modeling enables developers and security professionals to design systems that are inherently more resilient to attacks, incorporating security best practices and principles from the outset.

→ **Facilitation of Compliance with Regulations:** Threat modeling can also assist organizations in meeting various regulatory and compliance requirements. Many security standards and regulations mandate that organizations take proactive steps to identify and mitigate security risks. By conducting thorough threat modeling and documenting the process, organizations can demonstrate their commitment to security and their compliance with these requirements.

Threat Modeling Through the Lens of "The Water Book"

The core principles of "The Water Book" resonate strongly with the essential concepts of effective threat modeling, providing valuable insights into how to approach this critical security practice:

→ **Adaptability, Fluidity, and Dynamic Response:** Just as water adapts its shape to fit any container, threat modeling must be a flexible, adaptable, and dynamic process. It should not be a static or one-time activity, but rather an ongoing and iterative process that evolves as the system changes, new threats emerge, the attack landscape shifts, and the organization's security posture matures. Threat modeling must be able to adapt to new information, changing circumstances, and emerging challenges.

→ **Strategic Approach, Mindset, and Foresight:** Threat modeling is not simply a tactical exercise focused on identifying technical vulnerabilities; it requires a strategic and proactive approach, demanding a deep understanding of the attacker's mindset, their potential motivations, and the various attack vectors they might employ. It's about anticipating potential attacks, thinking like an adversary, and developing proactive defense strategies.

→ **Balance, Perspective, and Contextual Awareness:** Threat modeling necessitates a balanced and comprehensive perspective that takes into consideration not only the technical aspects of security but also the broader business context in which the system operates. It's about understanding the potential impact of identified threats on the organization's overall objectives, priorities, and risk tolerance. Threat modeling should be informed by a strong awareness of the business environment, the organization's risk appetite, and the potential consequences of security failures.

Key Elements and Methodologies in Threat Modeling

Effective threat modeling typically involves the following key elements and methodologies:

→ **Comprehensive System Understanding:** The foundation of any successful threat modeling effort is a thorough and detailed understanding of the system being analyzed. This includes a clear grasp of its architecture, design, functionality, components, data flows, dependencies, and interactions with other systems. Different modeling techniques, such as data flow diagrams (DFDs) and architectural diagrams, can be used to visualize and document the system.

→ **Systematic Threat Identification:** The next crucial step is to systematically identify potential threats that could affect the system. This involves considering a wide range of attack vectors, potential threat actors (both internal and external), their motivations, capabilities, and the tactics, techniques, and procedures (TTPs) they might employ. Threat intelligence and knowledge of common attack patterns are essential for this stage.

→ **Vulnerability Analysis and Exploitation Scenarios:** Once potential threats have been identified, the next step is to analyze potential vulnerabilities that could be exploited to carry out those threats. This involves a detailed examination of the system's weaknesses, flaws, and design or implementation errors that could be leveraged by attackers. Developing realistic exploitation scenarios helps to understand how attackers might chain together different vulnerabilities to achieve their objectives.

→ **Risk Assessment, Prioritization, and Impact Analysis:** The identified threats and vulnerabilities are then assessed and prioritized based on their likelihood of occurrence and their potential impact on the organization. This risk assessment process helps to determine which threats pose the greatest risk and require the most urgent attention. Factors such as the sensitivity of the data, the criticality of the system, and the potential financial, reputational, and operational consequences of a successful attack are   considered.

→ **Development and Implementation of Mitigation Strategies:** Finally, appropriate and effective mitigation strategies are developed and implemented to address the identified threats and vulnerabilities. These strategies may include technical controls (e.g., encryption, access controls, firewalls), process improvements, security awareness training, and other measures aimed at reducing the likelihood or impact of potential attacks. Mitigation strategies should be tailored to the specific risks and vulnerabilities identified during the threat modeling process.

Threat modeling is not just a technical exercise; it's a critical, proactive, and strategic security practice that aligns strongly with the principles of "The Water Book". By embracing adaptability, adopting a strategic and attacker-centric approach, and maintaining a balanced and contextually-aware perspective, organizations can leverage threat modeling to build more secure and resilient systems, prioritize their security efforts effectively, and proactively mitigate potential risks in an ever-changing and increasingly complex threat landscape.

Balancing Security and Business Needs: A Strategic Imperative

In the intricate and dynamic world of cybersecurity, one of the most crucial and perpetually challenging responsibilities of a cyber strategist is to achieve and maintain a delicate equilibrium between the organization's security imperatives and its overarching business needs. Security cannot and must not operate in isolation; it must be seamlessly integrated into the organization's daily operations and strategic objectives in a way that actively supports and facilitates its core mission, rather than impeding or hindering it. This necessitates a nuanced, comprehensive, and adaptive understanding of both the ever-evolving cybersecurity landscape and the organization's specific business priorities, coupled with highly effective communication and collaboration with a diverse range of stakeholders across the entire organization.

The Agile Approach, Strategic Thinking, and "The Water Book"

Earlier in our exploration, we emphasized the critical importance of adopting an agile approach to cybersecurity. We drew parallels to the fluid and adaptable nature of water, as described in "The Water Book." Just as water effortlessly adapts its form to fit any container, cybersecurity strategies and practices must be flexible, adaptable, and highly responsive to the business's constantly changing and often unpredictable needs. This agility is not merely beneficial; it's essential when it comes to effectively balancing security and business needs, ensuring that security measures are implemented in a way that supports and enhances – rather than disrupts – the organization's operations.

Key Strategies for Balancing Security and Business Needs

To navigate this complex challenge effectively, cyber strategists should employ a range of key strategies:

→ **Deep Understanding of Business Priorities:** A successful cyber strategist must possess a deep and thorough understanding of the organization's core business objectives, its primary revenue streams, its critical operational processes, and its overall risk tolerance. This understanding goes beyond superficial awareness; it requires a detailed and nuanced grasp of what drives the business, its priorities, and its risk appetite in different areas. Security measures should be carefully designed and implemented to minimize any disruption to critical business functions, ensure business continuity, and actively support the organization's ability to achieve its strategic goals and objectives.

→ **Effective and Persuasive Communication:** Highly effective communication is an essential skill for any cyber strategist seeking to balance security and business needs. Security professionals must be able to clearly and persuasively articulate security risks in terms that resonate with business leaders and other non-technical stakeholders. This involves translating complex and often technical cybersecurity jargon into clear, concise, and easily understandable language that highlights the potential impact of security breaches on the organization's bottom line, reputation, competitive advantage, and ability to achieve its strategic objectives. Communication effectively and building bridges between security and business are paramount.

→ **Tailoring Security to the Organization's Context:** There is no universal, one-size-fits-all approach to cybersecurity. The appropriate and effective level of security for an organization will vary significantly depending on a multitude of factors, including the organization's specific industry, its size and complexity, its unique risk profile, and its specific business priorities. A cyber strategist must be able to assess these factors and tailor security measures to fit the organization's unique context. The ultimate goal is to implement security measures that provide robust and adequate protection against relevant threats without being overly burdensome, restrictive, or detrimental to business operations.

→ **Embracing a Collaborative and Integrated Approach:** Balancing security and business needs necessitates a collaborative and integrated approach. Security professionals should proactively and consistently work closely with business leaders from various departments, IT teams, and other relevant stakeholders to ensure that security considerations are seamlessly integrated into business processes, workflows, and decision-making processes. This collaborative approach fosters a shared understanding of security risks, promotes a sense of shared responsibility for maintaining a strong security posture, and ensures that security becomes an enabler of business success, rather than an impediment.

→ **Adopting a Risk-Based Approach:** A risk-based approach is fundamental to balancing security and business needs. This approach involves identifying, assessing, and prioritizing security risks based on their potential impact on the organization's business objectives. By focusing on the risks that pose the greatest potential harm, organizations can allocate their resources effectively and implement security measures that provide the most significant return on investment in terms of risk reduction. This approach ensures that security efforts are aligned with business priorities and that security investments are justified by their potential to protect the organization's most valuable assets.

→ **Promoting Security Awareness and Culture:** Creating a strong security awareness and culture throughout the organization is crucial for balancing security and business needs. When employees at all levels understand the importance of security and their role in maintaining it, they are more likely to support security initiatives and follow security policies. This reduces the friction between security and business operations and fosters a sense of shared responsibility for security.

→ **Continuous Monitoring and Adaptation:** The threat landscape and business needs are constantly evolving. Therefore, it's essential to continuously monitor the effectiveness of security measures and adapt them as needed. This ensures that security remains aligned with business objectives and continues to provide adequate protection against emerging threats.

Risk Tolerance and Communication: A Cornerstone of Cyber Strategy

An organization's risk tolerance serves as a crucial compass, defining the level of risk it is willing to accept and embrace in the pursuit of its core business objectives and strategic goals. This risk tolerance is not a static or fixed value; rather, it's a dynamic and context-sensitive concept that varies depending on a multitude of factors, including the specific situation, the potential impact and consequences of a security event or incident, and the organization's overarching priorities and strategic direction. Therefore, the ability to effectively and clearly communicate this risk tolerance, both within the cybersecurity team and to a diverse range of stakeholders across the organization, is not merely a desirable skill; it's an absolutely critical and indispensable component of a successful and robust cybersecurity strategy.

The Foundational Importance of Clear and Consistent Communication

Clear and consistent communication about risk tolerance is of paramount importance for several fundamental reasons:

→ **Alignment with Business Objectives and Strategy**: When an organization's risk tolerance is clearly defined, articulated, and consistently communicated, it enables the cybersecurity team to align its security efforts and initiatives with the organization's broader business objectives and strategic direction. This alignment ensures that security measures are implemented and enforced in a way that actively supports and facilitates the business, rather than hindering or impeding its progress. It's about making security an enabler   of   business   success,   not   an   obstacle.

→ **Facilitation of Informed Decision-Making**: Clear and transparent communication about risk tolerance plays a vital role in enabling informed and effective decision-making at all levels of the organization, from the executive suite to individual departments. Business leaders and decision-makers need a clear understanding of the potential risks associated with various business decisions, initiatives,   and   projects.   Simultaneously,   cybersecurity professionals need a deep understanding of the organization's risk appetite and tolerance levels to make appropriate and well-informed recommendations regarding security measures and risk mitigation   strategies.

→ **Optimization of Resource Allocation:** A thorough understanding of risk tolerance empowers organizations to allocate their often-limited resources in the most effective and efficient manner. By focusing their attention and resources on the areas where the organization has the lowest tolerance for risk, cybersecurity teams can prioritize their efforts, optimize their resource allocation, and ensure that security investments are used strategically to provide the greatest possible protection for the organization's most critical assets   and   operations.

→ **Cultivation of Stakeholder Buy-in and Shared Responsibility:** When risk tolerance is communicated effectively, clearly, and consistently, it fosters a strong sense of buy-in and shared responsibility among stakeholders across the organization. This helps to cultivate a pervasive culture of security awareness, where every employee understands their individual role in managing and mitigating risk, and where security becomes a collective responsibility, rather than solely the domain of the security team.

Strategic Approaches for Effective Risk Tolerance and Communication

To ensure that risk tolerance is effectively understood, clearly communicated, and consistently applied throughout the organization, cyber strategists should consider and implement the following strategic approaches:

→ **Defining and Categorizing Risk Tolerance Levels:** Traditional approaches often categorize risk tolerance using subjective levels such as 'high,' 'medium,' and 'low.' However, these qualitative labels inherently lack the precision and objectivity required for effective strategic decision-making and clear communication with business leadership. A more robust and business-aligned methodology involves defining risk tolerance through Cyber Risk Quantification (CRQ). This approach translates the potential impact of security incidents into concrete financial terms, such as estimated monetary losses or ranges of probable loss exposure. By quantifying risk exposure in this manner, organizations can more accurately align their tolerance levels with core business objectives, financial stability, reputation, and strategic goals. This provides a data-driven foundation for making informed decisions regarding risk acceptance, justifying mitigation investments based on potential ROI, and developing an overall cybersecurity strategy that speaks the language of the business.

→ **Tailoring Communication to Specific Audiences:** Communication about risk tolerance should be thoughtfully tailored to the specific needs and understanding of each target audience. For instance, business leaders and executives may require a high-level overview of the organization's overall risk tolerance and its implications for strategic decision-making. In contrast, technical staff and security professionals may need more detailed and granular information about specific risk tolerances related to systems, applications, and data.

→ **Leveraging Visual Aids and Representations:** The use of visual aids, such as charts, graphs, and dashboards, can be highly effective in communicating risk tolerance levels and related information. These visual representations can make it easier for stakeholders to quickly grasp and understand the organization's risk appetite, the potential impact of security events, and the overall risk landscape.

→ **Establishing Regular Review and Update Processes:** Risk tolerance is not a static concept; it's dynamic and subject to change over time. Therefore, it's essential to establish processes for regularly reviewing and updating risk tolerance levels to reflect changes in the threat landscape, evolving business objectives, new regulatory requirements, or significant shifts in the organization's risk appetite.

→ **Fostering Open and Transparent Dialogue:** Encourage and promote open, transparent, and honest communication about risk tolerance at all levels of the organization. This involves creating an environment where stakeholders feel comfortable expressing their concerns, asking questions, and engaging in constructive discussions about risk management and security.

Building Resilience and Redundancy: A Strategic Imperative for Survival

Resilience refers to an organization's fundamental ability to anticipate, withstand, adapt to, and rapidly recover from disruptions, regardless of whether those disruptions are caused by malicious cyberattacks, unexpected system failures, human error, or other adverse events. Redundancy, which is a critical component and enabler of resilience, involves the strategic implementation of backup systems, redundant processes, or alternative capabilities that can seamlessly take over and maintain essential functions in the event of a failure, attack, or disruption. Together, resilience and redundancy constitute a critical and indispensable defense strategy that empowers organizations to minimize downtime, ensure business continuity, protect their critical assets, and maintain operational stability in the face of adversity.

The Strategic Importance of Resilience and Redundancy:

Building a strong foundation of resilience and redundancy is of paramount strategic importance for several compelling reasons:

→ **Minimizing Downtime and its Impact:** Cyberattacks, system failures, and other disruptive events can inflict significant downtime on an organization, which can lead to substantial financial losses, severe reputational damage, erosion of customer trust, and major operational disruptions. Redundancy plays a crucial role in minimizing downtime by providing readily available backup systems, processes, and capabilities that can be activated to take over critical functions seamlessly in the event of a failure or attack, ensuring that essential operations can continue with minimal   interruption.

→ **Ensuring Business Continuity in the Face of Adversity:** In today's interconnected and digital driven world, organizations rely heavily and often critically on their IT systems and infrastructure to conduct their core business operations. Resilience is essential for ensuring that these critical business functions can continue to operate effectively and efficiently, even when the organization is confronted with significant challenges, disruptions, or adverse events. It's about maintaining operational stability and ensuring that the business can continue to deliver its products or services to its customers without unacceptable delays or interruptions.

→ **Robust Protection of Critical Assets:** Redundancy provides a strong layer of protection for an organization's most critical assets, which include sensitive data, business-critical applications, and essential IT infrastructure components. By having backup copies of data and redundant systems in place, organizations can effectively restore operations and recover from data loss or system compromise, ensuring that their most valuable assets are safeguarded   against   potential   threats.

→ **Enhancing Recovery Capabilities and Effectiveness:** Resilience goes far beyond simply having backups or redundant systems; it encompasses having a well-defined, tested, and executable plan, along with the necessary capabilities, to recover swiftly, efficiently, and effectively from any disruption. This includes having clearly documented and regularly updated recovery procedures, conducting thorough testing of those procedures to ensure their efficacy, and having access to the necessary resources, both human and technological, to execute the recovery plan successfully.

Strategic Approaches for Building Resilience and Redundancy

To effectively build resilience and redundancy into their cybersecurity posture, organizations should strategically consider and implement the following key approaches:

→ **Strategic Implementation of Redundant Systems:** This involves the strategic deployment of backup servers, redundant network devices, and other critical infrastructure components that can seamlessly take over and maintain operations in the event of a system failure, hardware malfunction, or cyberattack. Redundancy should be built into the architecture of critical systems to minimize single points of failure and ensure high availability.

→ **Robust Data Backup and Recovery Processes in the Age of Ransomware:** Regular and reliable data backups are absolutely essential, forming the bedrock of resilience against data loss, corruption, or compromise, particularly from pervasive threats like ransomware. Organizations must establish and maintain meticulous data backup and recovery processes. This involves defining appropriate backup frequencies, clear data retention policies, and ensuring the secure storage of backups. Critically, since ransomware often targets backup repositories to prevent recovery, implementing immutable backups has become a vital security layer. Immutable backups are designed to be unchangeable and undeletable for a predetermined period, providing a trustworthy restore point even if primary systems and standard backups are compromised. A comprehensive and regularly tested data recovery plan is indispensable alongside robust backup strategies. This plan must clearly outline the specific steps needed to restore data and critical systems efficiently, ensuring a swift return to operational status following any disruption, including a sophisticated ransomware attack.

→ **Comprehensive Disaster Recovery Planning**: A comprehensive and well-articulated disaster recovery plan is a critical component of resilience. This plan should outline the specific steps an organization will take to recover from a major disruption, such as a large-scale cyberattack, a natural disaster, or a widespread pandemic. The disaster recovery plan should include detailed procedures for restoring systems, recovering data, relocating operations if necessary, and maintaining clear and consistent communication with all relevant stakeholders, including employees, customers, and partners.

→ **Well-Defined Incident Response Planning:** As discussed earlier in Key 2, Agility, a well-defined and regularly tested incident response plan is absolutely crucial for minimizing the impact of any cyberattack or security incident. This plan should clearly outline the specific steps involved in detecting, containing, eradicating, and recovering from an incident, as well as procedures for post-incident analysis and lessons learned.

→ **Regular Testing, Exercises, and Simulations:** Consistent and rigorous testing, exercises, and simulations are essential for validating the effectiveness of resilience and redundancy measures. This includes thorough testing of backup systems, detailed recovery procedures, and the incident response plan to ensure that they function as expected and that the organization is well-prepared to respond to and recover from disruptive events. These tests and exercises should be conducted regularly and should involve various scenarios to ensure that all aspects of the recovery process are validated.

Drawing Wisdom from "The Water Book"

The core principles of "The Water Book" can provide valuable insights into our approach to building resilience and redundancy. Just as water adapts to its environment, flows around obstacles, and finds alternative paths, resilient systems should be designed with the inherent ability to adapt to disruptions, maintain essential functionality, and continue operating effectively. Redundancy provides those essential alternative paths – ensuring that if one path is blocked or compromised, another path is readily available to maintain business operations and minimize disruption.

Building resilience and redundancy is not merely a tactical consideration; it's a strategic imperative and a core component of a robust and effective cybersecurity strategy. By strategically implementing redundant systems, establishing and maintaining robust data backup and recovery processes, developing comprehensive and well-articulated disaster recovery and incident response plans, and conducting regular and rigorous testing, exercises, and simulations, organizations can significantly enhance their ability to withstand, adapt to, and rapidly recover from a wide range of disruptions, ensuring business continuity, protecting their critical assets, and maintaining operational stability in an increasingly challenging and unpredictable cyber landscape.

Summary of Key 2: Agility

Our exploration has shown that it is about more than just quick reactions; it's a fundamental cybersecurity strategy approach emphasizing adaptability, resilience, and balance. It draws heavily from the philosophy of Miyamoto Musashi's "The Water Book," which uses water as a metaphor for achieving these qualities.

Here's a breakdown of how the concepts we discussed embody the spirit of "The Water Book:"

→ **Incident Response:** We discussed developing a plan, using the OODA Loop (Observe, Orient, Decide, Act) for rapid decision-making, and learning from incidents. Just as water adapts to its container, incident response requires flexibility and the ability to change course quickly. The OODA Loop itself is a cycle of adaptation, mirroring how water flows and adjusts.

→ **Adapting to Change:** Agile security frameworks, DevSecOps, cloud security, and threat modeling were discussed. This reflects the water-like property of being able to flow and adjust to new environments   and   challenges.

→ **Maintaining Balance:** Balancing security with business needs, understanding risk tolerance, and building resilience and redundancy are crucial. Water maintains balance and finds equilibrium, and a CISO must do the same to ensure security efforts support business objectives without hindering them. Resilience and redundancy ensure that even when "obstacles" arise (like a cyberattack), the organization can still function, much like water finding a new path.

In essence, Key 2 and its components embody the core tenets of "The Water Book:"

→ **Adaptability**: Just as water adapts to any container, cybersecurity strategies must adapt to ever-changing threats and business needs.

→ **Agility**: Water flows quickly and efficiently; cybersecurity requires rapid   response   and   nimble   adjustments.

→ **Resilience:** Water overcomes obstacles; cybersecurity must ensure continued   function   even   amid   disruptions.

→ **Balance:** Water finds equilibrium; cybersecurity must balance security with business needs and risk tolerance.

By embracing these principles, cybersecurity professionals can move beyond rigid, static defenses and adopt a more fluid, dynamic, and ultimately more effective approach to protecting their organizations.

Key 3: Action (The Fire Book)

"Victorious warriors win first and then go to war, while defeated warriors go to war first and then seek to win." — Sun Tzu

Musashi, in his infinite wisdom, titled this section "The Fire Book." Appropriate, don't you think? Because in the world of cybersecurity, action isn't just about putting out fires, although let's be honest, there's plenty of that. It's about proactively taking the fight to the adversary. It's about playing defense and understanding the art of offensive security. As a seasoned cyber strategist, I've always believed that a deep understanding of the enemy's tactics is crucial to our defense. It's like knowing your opponent's favorite move in a chess game; it gives you the upper hand.

This "Fire Book" section isn't just about reacting to incidents; it's a comprehensive guide to proactive and offensive strategies. We'll explore the nuances of penetration testing and red teaming, the critical importance of vulnerability remediation and patch management, and how security automation and orchestration can amplify our actions. Let's face it: In this digital age, speed and precision can often mean the difference between a secure system and a breached one.

Ever wonder how the attackers think? Our exploration into the attacker mindset and threat intelligence comes in here. We'll also discuss the ethical tightrope walk of ethical hacking and responsible disclosure. And because a good offense is the best defense, we'll delve into disrupting attacker operations and active defense strategies.

Finally, because no CISO is an island, we'll discuss collaboration and information sharing. After all, sharing is caring, right? Even in cybersecurity. So buckle up because this is where the rubber meets the road, or in our case, where the firewall meets the fire.

Proactive Defense

Penetration Testing and Red Teaming

In the intricate dance of cybersecurity, proactive defense isn't merely a strategic advantage; it's an existential imperative. It's the demarcation line between reactive scrambling and assertive control of our digital domain. To borrow Musashi's wisdom, it's about not just knowing the terrain, but dictating the terms of engagement. It's the cyber equivalent of a preemptive strike, but with less collateral damage and hopefully fewer international incidents.

Penetration testing forms a cornerstone of this proactive stance. Envision it as a controlled cybersecurity exercise, a simulated assault on our own fortifications. It's a systematic approach to evaluating the robustness of a computer system or network by meticulously emulating the tactics of malicious actors, both those lurking outside our digital walls and those who might have already slipped inside. This process involves a detailed and methodical analysis of the system, probing for vulnerabilities that could stem from a myriad of sources. These can include oversights in system configuration, those pesky latent hardware or software flaws that always seem to pop up at the worst time, or weaknesses in our operational or technical countermeasures. It's akin to putting a bridge through a rigorous stress test to ensure it can bear the weight of traffic, except in our world, the "traffic" is a relentless barrage of malicious cyber activity.

Red teaming elevates this proactive stance to a high art. It's penetration testing on an advanced, almost theatrical level. While penetration testing often focuses on specific vulnerabilities, red teaming adopts a more holistic and adversarial approach. It involves a cadre of seasoned security experts who immerse themselves in the roles of sophisticated adversaries, challenging an organization's security posture with the same cunning, creativity, and tenacity as real-world attackers. Red teams don't just look for the obvious entry points; they attempt to exploit every conceivable weakness, mimicking advanced persistent threats (APTs) to expose vulnerabilities that might otherwise remain hidden. They don't just check if the front door is locked; they try to pick the locks, climb through windows, and, if necessary, metaphorically tunnel underneath the building. Their arsenal includes the full spectrum of tactics, techniques, and procedures (TTPs) employed by actual attackers, making it a comprehensive and disturbingly realistic assessment of an organization's defenses.

Both penetration testing and red teaming are indispensable tools in a forward-thinking CISO's arsenal. They provide invaluable insights, illuminating an organization's security strengths and, more critically, exposing its weaknesses. They empower us to identify vulnerabilities before they can be exploited by malicious actors, granting us the opportunity to reinforce our defenses, strengthen our security posture, and ultimately, to sleep a little better at night. In essence, they transform us from reactive defenders, constantly putting out fires, into proactive guardians, anticipating and neutralizing threats before they materialize.

However, the discovery of vulnerabilities is merely the opening act. Vulnerability remediation and patch management are the crucial follow-through, the execution of our proactive strategy. A robust and well-defined process for identifying, assessing, and remediating vulnerabilities in a timely and efficient manner is paramount. It's not a one-time fix but a continuous cycle of improvement, a commitment to staying one step ahead of the ever-evolving adversary. Patch management, while sometimes relegated to the realm of compliance, is a critical element in this process. It's about more than just applying updates; it's about understanding the associated risks, prioritizing remediation efforts based on potential impact, and ensuring that your systems are as fortified as possible against known exploits.

In today's increasingly complex and interconnected digital landscape, security automation and orchestration are no longer luxuries or buzzwords; they're absolute necessities. Security teams often grapple with a deluge of alerts, a tidal wave of data, and a shortage of resources, making it incredibly challenging to respond to threats in a timely and effective manner. Security automation and orchestration streamline security operations, automating repetitive and mundane tasks, freeing up valuable human resources, and significantly improving incident response times. It's akin to having a well-coordinated and highly efficient team of cybersecurity professionals working in perfect harmony, responding to threats with machine-like speed and precision. Think of it as a finely tuned orchestra, where each instrument plays its part seamlessly, guided by the conductor's baton.

In the final analysis, proactive defense is not just about having the latest and greatest tools or technologies. It's fundamentally about a mindset, a philosophy, a way of approaching cybersecurity. It's about embracing a proactive stance, understanding the adversary's tactics and motivations, and continuously striving to improve our defenses. It's about embodying the spirit of Musashi's "Fire Book" and taking the fight to the enemy, not waiting for them to bring the fight to us. It's about being the chess player who anticipates their opponent's moves several steps ahead, not the one who only reacts to the immediate threat. It's about being the cyber strategist, not just a cyber technician.

Vulnerability Remediation and Patch Management

In the intricate landscape of cybersecurity strategy, vulnerability remediation and patch management stand as indispensable disciplines. It's not merely a task to be completed; it's a continuous commitment to maintaining the integrity and resilience of an organization's digital infrastructure. This process demands a systematic, rigorous, and proactive approach to identifying, assessing, and mitigating vulnerabilities within systems and networks. It's a cycle of perpetual improvement, a dedication to staying ahead of the ever-evolving threat landscape. As the cybersecurity luminary Bruce Schneier astutely observed, "Security is not a product, but a process," a principle that underscores the dynamic and ongoing nature of vulnerability management.

Third-Party Risk Management (TPRM) introduces significant complexities to this already challenging endeavor. In today's interconnected digital ecosystem, organizations increasingly rely on external vendors for various services and functionalities. While these partnerships can bring numerous benefits, they also expand the attack surface and introduce potential vulnerabilities that lie beyond an organization's direct control. The data paints a stark and concerning picture: "In 2033, 63 attacks on vendors: from those 63 attacks, 298 data breaches occurred across impacted companies," as reported by Black Kite. This sobering statistic underscores the urgent and critical need for robust and effective TPRM strategies. Traditional qualitative risk scoring methods have proven woefully inadequate in addressing these multifaceted challenges. These methods, often subjective, prone to bias, and lacking in data-driven insights, fail to provide the level of granular visibility and actionable intelligence necessary for proactive and informed decision-making.

The interconnectedness of modern businesses amplifies the challenge, with reliance on numerous vendors, cloud services, and third-party software. This web of dependencies means a vulnerability in one supplier's software can rapidly cascade across multiple organizations. In fact, the Black Kite 2025 Supply Chain Vulnerability Report reveals the sheer volume of vulnerabilities continues to surge, with over 40,000 Common Vulnerabilities and Exposures (CVEs) published in 2024 alone, a 38% year-over-year increase. This increase underscores the growing complexity of vulnerability management in the supply chain.

To navigate the intricate complexities of TPRM effectively and mitigate the risks posed by third-party vulnerabilities, a fundamental paradigm shift is necessary. Organizations must move decisively away from subjective assessments and embrace quantitative risk assessment methods, leveraging cutting-edge technologies such as artificial intelligence (AI) and machine learning (ML) to automate and enhance risk identification and mitigation processes. Agility becomes paramount in this new paradigm, enabling organizations to mount rapid and effective responses to emerging threats and adapt dynamically to the ever-shifting threat landscape. The OODA loop (Observe, Orient, Decide, Act), a concept originally conceived by military strategist John Boyd and later adapted for various fields, including cybersecurity, provides a valuable framework for this agile approach. It emphasizes the importance of continuous observation of the environment, orientation to understand the implications of observations, rapid decision-making, and decisive action, all within a cyclical and iterative process that allows for swift adaptation to changing circumstances.

Cyber Risk Quantification (CRQ) plays an increasingly crucial role in this transformative process. It represents a significant departure from subjective assessments by translating the often-nebulous realm of cyber risk into quantifiable metrics, enabling clearer communication of risk to stakeholders and facilitating more informed and data-driven decision-making. Models such as the Factor Analysis of Information Risk (FAIR) provide standardized taxonomies and frameworks for this purpose, allowing organizations to calculate the probable financial impact of potential cyber events and align cybersecurity initiatives more effectively with overarching business goals. As Jack Jones, co-founder of the FAIR Institute, astutely asserts, "You can't manage what you can't measure," highlighting the fundamental importance of quantitative risk assessment in effective cybersecurity management.

The integration of AI and machine learning further enhances the capabilities of vulnerability remediation and patch management. These transformative technologies can analyze vast datasets with unparalleled speed and efficiency, detect patterns and anomalies that might escape human observation, and automate the identification and mitigation of potential threats, thereby streamlining TPRM processes, improving risk prediction, and strengthening overall cybersecurity posture.

For example, AI algorithms can automate the process of vulnerability detection in third-party systems, significantly reducing the time and resources required for this critical task. ML models, trained on historical data, can predict potential threats and assist in proactive risk management, enabling organizations to anticipate and mitigate risks before they materialize. However, it is crucial to acknowledge and address the challenges associated with the implementation of AI and machine learning in this context, including the need for significant investment in infrastructure and specialized skills, as well as the potential for algorithmic bias and the dependence on the quality and integrity of training data.

Vulnerability remediation and patch management are not isolated tasks but fundamental components of a comprehensive and proactive cybersecurity strategy. Embracing agility, leveraging the precision of cyber risk quantification, and harnessing the transformative capabilities of AI and machine learning are essential imperatives for organizations seeking to navigate the increasing complexities of modern cyber threats and effectively manage the inherent risks associated with third-party relationships. Adopting this stance requires a decisive shift towards a more dynamic, data-driven, and automated approach, recognizing that in our interconnected digital world, an organization's security posture is often inextricably linked to the strength and resilience of its weakest link.

Security Automation and Orchestration

Security automation and orchestration have transcended the realm of mere buzzwords; they have become indispensable cornerstones of a robust and effective cybersecurity strategy.

We're not simply talking about automating mundane tasks; we're discussing the creation of a dynamic and interconnected cybersecurity ecosystem capable of operating with unparalleled speed, precision, and efficiency. It's about transitioning from a reactive security posture, constantly playing catch-up with threats, to a proactive and even predictive stance, anticipating and neutralizing threats before they can inflict damage. Think of it as moving from a manual approach to cybersecurity, where human operators are constantly required to intervene, to an automated system that can operate autonomously, much like shifting from manually piloting an aircraft to utilizing an autopilot system that can maintain course and respond to changes in the environment with minimal human input.

The imperative for security automation and orchestration becomes even more pronounced and critical when we consider the accelerating rise of Agentic AI. Agentic AI, characterized by its ability to perform tasks autonomously, make decisions, and even learn and adapt, presents both immense opportunities and significant challenges for the field of cybersecurity. On one hand, Agentic AI can be a formidable force multiplier, a powerful tool for automating a wide range of security tasks, detecting anomalies with greater precision and speed than human analysts, and responding to threats in real-time with minimal human intervention. It can act as a tireless and vigilant sentinel, constantly monitoring systems, analyzing data, and taking appropriate actions to protect against cyberattacks. On the other hand, this same technology also introduces new and complex attack vectors, significantly amplifying the potential impact and scale of cyberattacks.

Consider this: Agentic AI could be leveraged by security professionals to automate penetration testing processes, identifying vulnerabilities with greater speed, accuracy, and comprehensiveness than traditional human testers. It could also be used to orchestrate complex and coordinated security responses, seamlessly integrating and coordinating various security tools and systems to neutralize threats effectively and efficiently. Imagine an AI-powered system that can not only detect a zero-day vulnerability but also automatically trigger a series of actions, such as isolating affected systems, deploying patches, and notifying relevant personnel, all within a matter of seconds.

However, the same technology could be weaponized by malicious actors to launch sophisticated and highly targeted attacks, automate the spread of malware with unprecedented speed and reach, or even manipulate security systems and defenses, turning them against their intended purpose. Think of AI-powered malware that can learn and adapt to evade detection, or AI-driven phishing campaigns that can craft highly personalized and convincing messages to trick unsuspecting users. It's a double-edged sword, a powerful tool that can be used for both good and ill, much like Musashi's own weapon of choice – the katana – which could be used for both defense and offense.

Security automation and orchestration, particularly in the context of Agentic AI, must be approached with both enthusiasm and caution, with both a sense of opportunity and a healthy dose of skepticism. The balance is about strategically leveraging the immense power of automation to enhance our security capabilities, improve our efficiency, and strengthen our defenses, while simultaneously and proactively mitigating the inherent risks associated with its implementation and potential misuse. Accomplishing this requires a strategic and thoughtful approach, focusing on building resilient, adaptable, and intelligent systems that can learn, adapt, and evolve alongside the ever-changing and increasingly complex threat landscape. It's not just about automating tasks for the sake of automation; it's about automating intelligently and strategically, with a clear understanding of the potential benefits and risks involved. It's about building cybersecurity systems that can not only respond to threats but also anticipate them, learn from past experiences, and adapt to future challenges, much like a skilled warrior who can anticipate their opponent's moves and adjust their strategy accordingly. And let's be honest – in the rapidly evolving age of Agentic AI, where the lines between offense and defense are becoming increasingly blurred – that's not just a good idea; it's an absolute necessity for survival in the digital arena.

Side Note

I believe that Security Automation and Orchestration (SAO) is crucial for modern cybersecurity, enhancing efficiency and incident response, and the integration of Generative and Agentic AI offers transformative potential by automating threat intelligence, generating security content, enabling autonomous threat detection, and facilitating intelligent responses. At the same time, this synergy promises significant improvements in security posture, it also presents challenges in integration, the need for skilled personnel, and the risk of over-automation, necessitating a careful and strategic approach to implementation and continuous adaptation, which will ultimately reshape the roles of security professionals in an AI-augmented Security Operations Center. To get a deeper understanding of the current state, I asked Gemini Deep Research to conduct a review, the results of which are included in Appendix C.

Offensive Security

Understanding the Attacker Mindset

A purely defensive posture is akin to a warrior who has mastered the art of blocking but remains oblivious to the nuances of their opponent's fighting style. To truly fortify our digital domains, we must transcend the limitations of reactive security and embrace the proactive discipline of offensive security, which begins with a deep and nuanced comprehension of the attacker mindset. We must embody the strategic wisdom of masters like Musashi, who understood that victory is often determined not by brute force but by an acute awareness of the enemy's intentions, their preferred tactics, and their underlying motivations, often before the clash of swords even begins. This ancient wisdom translates directly into the modern context of cybersecurity: to defend effectively and strategically, we must think like an attacker, anticipate their moves, and unravel the complexities of their thought processes.

Understanding the attacker mindset involves a comprehensive and multifaceted exploration of their motivations, their tactical approaches, and their ultimate objectives. Attackers are not mere digital specters lurking in the shadows of the internet; they are individuals or groups with specific and often well-defined goals and strategies that meticulously guide their actions. Some are driven primarily by the pursuit of financial gain, seeking to exploit vulnerabilities in systems and networks to steal sensitive data for lucrative profit, extort organizations through ransomware attacks, or disrupt critical operations to demand ransom payments. The Black Kite report, for instance, poignantly illustrates the significant financial risks that organizations face due to third-party breaches, underscoring the substantial monetary rewards that attackers often seek. To illustrate, consider the devastating impact of ransomware attacks on healthcare providers, where attackers encrypt patient data and demand exorbitant ransoms, jeopardizing patient care and inflicting significant financial losses.

Others may be motivated by ideological or political convictions, seeking to inflict damage, cause disruption, or spread propaganda against specific organizations, governments, or even entire societies. Nation-state actors, for example, may engage in cyber espionage to steal intellectual property or disrupt critical infrastructure as part of a broader geopolitical strategy. And then there are those attackers who are driven by the allure of the intellectual challenge, viewing cybersecurity as an intricate game of wits, seeking to test the boundaries of their skills and abilities against a worthy adversary – often with little regard for the consequences of their actions. Hacktivists, for instance, may launch cyberattacks to protest against perceived injustices or to promote specific political or social agendas. Each of these diverse motivations shapes the attacker's approach, influences their selection of targets, and determines the intensity and persistence of their efforts.

To effectively counter these varied and evolving threats, we must also meticulously analyze and dissect the tactics that attackers commonly employ. They may utilize sophisticated social engineering techniques to manipulate unsuspecting individuals into divulging sensitive information, exploiting human psychology and trust to bypass even the most robust technical defenses.

Attackers may also exploit software vulnerabilities, leveraging weaknesses in code, applications, or operating systems to gain unauthorized access to systems and networks. The Equifax breach serves as a stark reminder of the devastating consequences that can arise from failing to patch known vulnerabilities in a timely manner. Malware, malicious software designed to infiltrate and inflict damage upon computer systems, is another ubiquitous tool in the attacker's arsenal. Ransomware attacks, where attackers encrypt a victim's valuable data and demand a ransom payment in exchange for its release, have become increasingly prevalent in recent years, causing widespread disruption, significant financial losses, and reputational damage to organizations across the globe.

The attacker mindset is not a static and easily-defined entity; it's a dynamic, adaptive, and constantly evolving landscape, mirroring the rapid pace of technological advancements and the ingenuity of human malevolence. Attackers are perpetually adapting to new technologies, discovering novel vulnerabilities, and devising innovative and sophisticated attack methods to circumvent even the most advanced security measures.

To maintain a proactive security posture, anticipate emerging threats, and stay one step ahead of these adversaries, we must cultivate a mindset of continuous learning, relentless adaptation, and strategic thinking. We must proactively seek to improve our defenses, challenge established norms, and constantly question whether our current security measures are sufficient to withstand our adversaries' evolving tactics and strategies. It's a relentless and ongoing battle of wits, demanding a deep understanding of cybersecurity's technical intricacies and the often-unpredictable complexities of human behavior.

Threat Intelligence and Hunting

In cybersecurity's dynamic and often adversarial realm, a purely reactive approach is akin to a warrior who only reacts to the enemy's blows, never anticipating their movements or seeking to disrupt their plans. To truly protect our digital assets and maintain a proactive security posture, we must embrace threat intelligence and threat hunting disciplines. These practices are not merely about responding to incidents after they occur; they are also about actively seeking out potential threats, anticipating their arrival, and disrupting their operations before they can inflict damage. By embodying the strategic wisdom of Sun Tzu, who emphasized the importance of knowing your enemy and knowing yourself and applying that knowledge, you can gain a decisive advantage in the cyber battlefield.

Threat intelligence is, at its core, the process of collecting, analyzing, and disseminating actionable information about potential or current attacks. It's about developing a deep understanding of who the attackers are, what their motivations and capabilities are, and what tactics, techniques, and procedures (TTPs) they commonly employ. High-quality and relevant threat intelligence provides critical context, enabling organizations to make informed and strategic decisions about their security posture, allocate resources effectively, and prioritize their defensive efforts. Think of the skilled scout who ventures into enemy territory, gathers detailed reports about their movements, strengths, and weaknesses, and then delivers that intelligence to the army's leadership, allowing them to prepare and strategize accordingly.

Threat intelligence can be derived from a diverse range of sources, both internal and external to the organization. Internal sources might include system logs, network traffic data, security alerts generated by various security tools, and incident response reports. These internal sources provide valuable insights into past attacks and ongoing activity within the organization's environment. External sources, on the other hand, such as Black Kite's research emphasizes the importance of understanding the threat landscape and anticipating potential risks, particularly those associated with third-party vendors. This focus on third-party risk highlights the need to gather intelligence on the security posture of vendors and proactively hunt for vulnerabilities in their systems, demonstrating how threat intelligence extends beyond an organization's own boundaries and encompasses the entire digital ecosystem. The key is to gather relevant and reliable information from these various sources, correlate and analyze it effectively, and then disseminate it to the appropriate stakeholders within the organization to enable informed decision-making and proactive action.

Threat hunting takes this proactive approach a significant step further. It involves actively and aggressively searching for malicious activity that might be lurking within an organization's network or systems, rather than simply relying on automated alerts and reactive security monitoring. Unlike traditional security monitoring, which primarily focuses on detecting known threats and responding to predefined alerts, threat hunting is a more hands-on, investigative, and hypothesis-driven approach. Threat hunters are skilled and experienced security analysts who leverage their deep knowledge of attacker tactics, techniques, and procedures to actively search for indicators of compromise (IOCs) and other evidence of malicious activity that might have been missed or evaded by automated security systems. It's akin to having a team of highly skilled detectives who are actively searching for clues, following hunches, and pursuing leads, rather than passively waiting for a crime to be reported.

Threat hunting requires a combination of technical expertise, analytical skills, and a deep understanding of the attacker mindset. It often involves using a variety of tools and techniques, including advanced analytics, behavioral analysis, and anomaly detection, to identify suspicious activity that might not trigger traditional security alerts. Threat hunters must be able to think like attackers, anticipate their moves, and understand the subtle indicators that might reveal their presence within the network. They must also be able to effectively communicate their findings to other security teams and stakeholders, enabling them to take appropriate action to contain and remediate any threats that are discovered. In essence, threat intelligence provides the knowledge and context, while threat hunting puts that knowledge into action, proactively seeking out and neutralizing threats before they can cause significant harm.

Ethical Hacking and Responsible Disclosure

Ethical hacking, at its core, is the practice of using hacking techniques for defensive purposes. It's about employing the same tools and techniques that malicious attackers use, but with the explicit permission of the system owner and with the intention of improving security. Ethical hackers are essentially "white hat" hackers who work to find vulnerabilities in systems before the "black hat" hackers do. They're like security consultants who get paid to break into your house, but only to tell you where the weak spots are so you can fix them.

Ethical hacking involves a variety of techniques, including penetration testing, vulnerability scanning, and social engineering. Penetration testing, as we discussed earlier, is a controlled attack on a system to identify vulnerabilities. Vulnerability scanning uses automated tools to identify known weaknesses in systems. Social engineering testing evaluates an organization's susceptibility to social engineering attacks, such as phishing or pretexting.

Responsible disclosure is the practice of reporting vulnerabilities to the affected vendor or organization in a way that allows them time to fix the issue before it's publicly revealed. This allows the organization to address the vulnerability before malicious actors can exploit it. In this way, responsible disclosure is a crucial part of ethical hacking, as it ensures that vulnerabilities are addressed to minimize risk to users.

The concept of responsible disclosure is often contrasted with full disclosure, where vulnerabilities are publicly revealed immediately, regardless of whether a patch is available. While full disclosure can sometimes be used to pressure vendors to fix vulnerabilities quickly, it also carries the risk of exposing users to harm if the vulnerability is exploited before a patch is released. Responsible disclosure strikes a balance between these two extremes, prioritizing the security of users while also holding vendors accountable for addressing vulnerabilities.

Ethical hacking and responsible disclosure are crucial for maintaining a secure digital environment. They allow organizations to proactively identify and address vulnerabilities, reducing the risk of cyberattacks and data breaches. It's about using the same weapons as the enemy, but using them for good.

Taking the Fight to the Enemy

Disrupting Attacker Operations

The era of passive defense is waning. While fortifying our digital borders remains crucial, true security in the modern threat landscape demands a more assertive stance: taking the fight directly to the enemy by actively disrupting their operations. This isn't about reckless retaliation; it's about strategically dismantling their capabilities, hindering their progress, and ultimately making the cost of attack outweigh the potential gain.

Disrupting attacker operations requires a fundamental shift in mindset. We must move beyond simply reacting to attacks and embrace a proactive approach that seeks to dismantle the attackers' infrastructure, methodologies, and even their organizational structures. This involves several key elements:

1. Intelligence-Driven Disruption: The foundation of any successful disruption strategy lies in deep, actionable intelligence. We need to understand our adversaries – their motivations, their tools, their tactics, and their infrastructure. This intelligence allows us to identify key points of vulnerability within their operations and target them effectively. Imagine proactively identifying and neutralizing command and control servers before they can be used to launch attacks, or mapping out the network of a ransomware gang to target their key facilitators.

2. Active Deception and Misdirection on a Grand Scale: Honeypots and deception technologies have their place, but we need to think bigger. Can we create entire fabricated digital environments designed to lure attackers, waste their resources, and provide us with invaluable intelligence? Picture a scenario where attackers believe they have successfully infiltrated a high-value target, only to find themselves navigating a meticulously crafted illusion that allows us to study their techniques and even feed them disinformation.

3. Targeting the Attack Lifecycle: Disruption shouldn't wait until an attack is in full swing. We need to identify and target attackers at every stage of their lifecycle, from initial reconnaissance and weaponization to delivery and exploitation. This could involve proactively identifying and reporting malicious infrastructure, disrupting phishing campaigns before they reach their targets, or even working to dismantle underground marketplaces where attack tools and stolen data are traded.

4. Strategic Use of Countermeasures: While outright offensive cyber operations by private entities remain a complex and often legally ambiguous area, there are strategic countermeasures that can be employed to disrupt attackers. This could involve techniques like sink holing malicious domains, disrupting botnet command and control channels, or even deploying sophisticated network traffic analysis to identify and isolate malicious actors within our own infrastructure.

5. Collaboration and Information Sharing as a Weapon: Attackers thrive in the shadows, exploiting the lack of coordination and information sharing among potential targets. By fostering robust information sharing platforms and collaborative defense strategies, we can collectively identify and disrupt attacker operations more effectively. Imagine a global network of security professionals sharing real-time intelligence on emerging threats and coordinating efforts to neutralize them.

6. Legal and Law Enforcement Partnerships: Disrupting sophisticated attacker operations often requires the authority and resources of law enforcement. Building strong partnerships with legal authorities and providing them with the necessary intelligence and technical expertise is crucial for bringing cybercriminals to justice and dismantling their organizations.

Taking the fight to the enemy is not about engaging in digital warfare in the traditional sense. It's about adopting a proactive, intelligence-driven, and collaborative approach to disrupt attacker operations at every possible stage. It's about shifting the power dynamic and making the digital realm a far less hospitable environment for malicious actors. This is the essence of a truly resilient and secure digital future.

Side Note: *When preparing for this section, I conducted deep research using Gemini Advanced. The results of that research are in Appendix D.*

Active Defense: Shaping the Battlefield to Our Advantage

The shift from passive defense to actively engaging our adversaries marks a critical evolution in cybersecurity. Active defense isn't about simply reacting to attacks; it's about proactively shaping the digital battlefield to our advantage, disrupting attacker operations before they can achieve their objectives, and ultimately making our environments a hostile territory for malicious actors.

At its core, active defense is a philosophy and a set of strategies that go beyond traditional preventative and detective measures. It involves actively seeking out, engaging with, and neutralizing threats in a dynamic and adaptive manner. This requires a blend of sophisticated technologies, skilled personnel, and a proactive security posture.

Here are some key strategies that fall under the umbrella of active defense for disrupting attacker operations:

1. Advanced Threat Hunting: This isn't just about reacting to alerts; it's about proactively searching for indicators of compromise (IOCs) and indicators of attack (IOAs) that may have bypassed traditional security controls. Skilled threat hunters leverage deep knowledge of attacker tactics, techniques, and procedures (TTPs) to uncover hidden threats and disrupt them before they can escalate into full-blown incidents. This proactive approach allows us to identify and eliminate attackers who may be lurking within our networks.

2. Deception Technologies on Steroids: Moving beyond basic honeypots, advanced deception platforms create intricate and realistic decoy environments that mimic our production systems. These decoys are designed to lure attackers away from real assets, provide early warning of intrusions, and allow us to observe their techniques in a controlled setting. By wasting their time and resources on false targets, we actively disrupt their progress and gain valuable intelligence.

3. Dynamic Cyber Deception and Camouflage: This involves actively changing our digital footprint and presenting a constantly shifting landscape to potential attackers. This could include techniques like dynamic IP address allocation, randomized hostnames, and the creation of ephemeral systems that appear and disappear, making it significantly harder for attackers to map our infrastructure and maintain a foothold.

4. Incident Response as an Offensive Maneuver: When an incident occurs, active defense dictates that we don't just contain and eradicate the threat; we also actively gather intelligence about the attacker, their tools, and their objectives. This information can then be used to proactively hunt for other instances of the same attacker within our environment and even contribute to broader intelligence sharing efforts to disrupt their operations elsewhere.

5. Controlled Countermeasures (with Extreme Caution): In specific and carefully considered scenarios, active defense might involve deploying controlled countermeasures against identified attackers. This defense could range from blocking malicious traffic at its source to – in very limited and legally justifiable circumstances – disrupting their infrastructure. However, this area requires extreme caution, clear legal frameworks, and a deep understanding of potential unintended consequences. Misattribution or escalation can have severe repercussions.

6. Intelligence-Led Disruption Campaigns: Leveraging the intelligence gathered through threat hunting, deception, and incident response, organizations can launch targeted disruption campaigns against known threat actors. This could involve working with ISPs and hosting providers to take down malicious infrastructure, reporting attacker accounts to relevant platforms, or contributing to law enforcement investigations.

7. Proactive Vulnerability Management and Hardening: While seemingly traditional, an active defense approach to vulnerability management involves aggressively identifying and patching vulnerabilities before they can be exploited. This includes proactive threat modeling and simulating attacks to identify weaknesses in our defenses and address them before real attackers can.

The Benefits and Challenges

Active defense offers significant benefits in disrupting attacker operations, including early detection, reduced impact of attacks, enhanced intelligence gathering, and potential deterrent effects. However, it also presents challenges. It requires highly skilled security personnel, sophisticated technologies, and a significant investment in resources. Furthermore, the implementation of specific active defense strategies, particularly those involving countermeasures, carries inherent risks and requires careful consideration of legal and ethical implications.

Moving Forward

Active defense is not a one-size-fits-all solution. Organizations need to carefully assess their risk profile, resources, and legal constraints before implementing specific strategies. However, in the face of increasingly sophisticated and persistent threats, embracing a proactive and assertive security posture through active defense is becoming a necessity for those who truly want to take the fight to the enemy and disrupt their operations. It's about shifting from a purely reactive stance to one where we actively shape the battlefield and make it a dangerous and unprofitable place for cyber adversaries.

The Unbreakable Chain: Amplifying Disruption Through Collaboration and Information Sharing

The asymmetry of the conflict often favors the attacker. They operate with stealth, leveraging anonymity and distributed infrastructure, while defenders are usually siloed, protecting individual perimeters. We must forge an unbreakable chain of collaboration and information sharing to truly take the fight to the enemy and effectively disrupt their operations. This collective defense transcends organizational boundaries and transforms isolated pockets of knowledge into a unified force.

The power of collective defense lies in dismantling the attacker's advantage of operating in the shadows. When organizations and individuals openly and effectively share threat intelligence, best practices, and incident response experiences, we create a dynamic and adaptive security ecosystem that is far more resilient and capable of proactively hindering malicious actors. This isn't merely about exchanging data points; it's about building a shared understanding of the threat landscape, anticipating attacker movements, and orchestrating coordinated actions to dismantle their capabilities.

Consider the analogy of a network of interconnected sensors. Each individual sensor might detect a small anomaly, but when the data from multiple sensors is aggregated and analyzed, a clear picture of a larger threat emerges. Similarly, individual organizations might detect isolated attacks or suspicious activity, but when this information is shared and correlated across a wider community, we can identify broader campaigns, understand attacker methodologies, and pinpoint the infrastructure they rely upon.

Deepening the Collaboration Landscape

Collaboration in cybersecurity manifests in various forms, each contributing uniquely to the disruption effort:

Industry-Specific Information Sharing and Analysis Centers (ISACs): These sector-specific organizations facilitate the sharing of threat information and best practices among companies within the same industry. This targeted approach allows for the rapid dissemination of relevant intelligence about threats specifically targeting financial institutions, healthcare providers, energy companies, and other critical infrastructure sectors. ISACs often play a crucial role in coordinating incident response efforts and disseminating actionable mitigation strategies.

Public-Private Partnerships: Collaboration between government agencies and private sector organizations is vital for addressing national-level cyber threats. Governments possess unique intelligence capabilities and legal authorities, while the private sector often has firsthand experience dealing with the latest attack techniques. Effective partnerships enable the sharing of classified threat information, joint investigations, and the development of national cybersecurity strategies aimed at disrupting major threat actors.

Informal Peer-to-Peer Sharing: Beyond formal structures, informal networks of security professionals sharing information through online forums, social media groups, and personal connections can be incredibly valuable. These channels often facilitate the rapid dissemination of emerging threat intelligence and practical advice on dealing with new attacks.

Threat Intelligence Platforms and Feeds: Commercial and open-source threat intelligence platforms aggregate and curate threat data from various sources, providing organizations with actionable insights into emerging threats and attacker infrastructure. Sharing data through these platforms expands the reach of individual intelligence efforts and contributes to a more comprehensive understanding of the global threat landscape.

Joint Exercises and Simulations: Collaborative cybersecurity exercises and simulations allow organizations to test their incident response plans and identify areas for improvement in a realistic environment. These exercises also foster relationships and build trust among participating teams, which is crucial for effective collaboration during real-world incidents.

Illustrating Disruption Through Shared Knowledge

The impact of collaboration and information sharing on disrupting attacker operations is tangible:

→ **Ransomware Disruption:** When multiple organizations share information about specific ransomware variants, their command and control infrastructure, and the tactics employed by the attackers, security vendors and law enforcement agencies can develop decryption tools, track down the perpetrators, and dismantle their operations.

→ **Supply Chain Attack Mitigation:** Sharing intelligence about compromised suppliers or malicious updates allows downstream victims to proactively identify and mitigate the impact of supply chain attacks, preventing widespread damage.

→ **Nation-State Actor Tracking and Countermeasures:** Collaborative efforts involving government agencies, research institutions, and private sector companies can help to identify and track the activities of nation-state-sponsored threat actors, leading to the development of targeted countermeasures and diplomatic pressure.

→ **Botnet Takedowns:** Coordinated efforts involving internet service providers, security companies, and law enforcement agencies, fueled by shared intelligence about botnet infrastructure and command patterns, have led to the successful takedown of numerous large-scale botnets used for various malicious purposes.

Addressing the Hurdles to Seamless Collaboration

While the benefits are clear, several challenges can hinder effective collaboration and information sharing:

Trust Deficits: Organizations may be hesitant to share sensitive information due to concerns about reputational damage, legal liabilities, or the potential for the shared data to be misused. Building trust requires establishing clear guidelines, anonymizing sensitive information where appropriate, and fostering a culture of transparency and reciprocity.

Competitive Concerns: In some industries, organizations may view threat intelligence as a competitive advantage and be reluctant to share it with rivals. Overcoming this requires a shift in perspective, recognizing that collective security ultimately benefits everyone.

Legal and Regulatory Restrictions: Data privacy regulations and other legal frameworks can sometimes create complexities around sharing threat information across different jurisdictions. Navigating these regulations requires careful consideration and the development of clear legal guidelines for information sharing.

Lack of Standardization: Inconsistent data formats and sharing mechanisms can make it difficult to effectively aggregate and analyze threat intelligence from different sources. Adopting common standards and promoting interoperability are crucial for streamlining the sharing process.

Information Overload and Noise: The sheer volume of threat data can be overwhelming, making it difficult to identify the most relevant and actionable information. Effective collaboration requires mechanisms for filtering, prioritizing, and contextualizing shared intelligence.

Strategies for Enhancing Collaborative Defense

To truly harness the power of collaboration for disrupting attacker operations, we need to implement strategies that address these challenges:

Foster Trusted Communities: Invest in building and nurturing trusted networks for information sharing, both within specific industries and across broader cybersecurity communities.

Develop Clear Legal Frameworks and Safe Harbors: Governments should work to establish clear legal frameworks and safe harbor provisions that encourage responsible information sharing without fear of undue liability.

Promote the Adoption of Standardized Data Formats: Encourage the adoption of common data formats and sharing protocols to improve the interoperability of threat intelligence.

Invest in Technology for Secure and Efficient Sharing: Develop and deploy secure platforms and automated systems that facilitate the rapid and efficient exchange of threat information.

Cultivate a Culture of Sharing and Reciprocity: Encourage organizations to actively contribute to information sharing initiatives and recognize the collective benefits of doing so.

Support the Role of Threat Intelligence Platforms: Promote the use of reputable threat intelligence platforms and encourage organizations to contribute their own findings to these platforms.

Organize and Participate in Joint Exercises: Regularly participate in collaborative cybersecurity exercises to build relationships, test incident response capabilities, and improve coordination.

Looking Ahead: The Future of Collaborative Disruption

As the cyber threat landscape continues to evolve in complexity and sophistication, the importance of collaboration and information sharing will only continue to grow. We can expect to see even more sophisticated collaborative defense mechanisms emerge, potentially leveraging advancements in artificial intelligence and machine learning to automate the analysis and sharing of threat intelligence at unprecedented speeds and scales. The future of effectively taking the fight to the enemy and disrupting their operations hinges on our ability to build stronger, more resilient, and more interconnected cybersecurity communities. By working together, we can forge an unbreakable chain that will ultimately make the digital realm a far less hospitable place for those who seek to do harm.

Key 3: Embracing the Fire of Proactive Action

This chapter, aptly titled "Key 3: Action" in homage to Musashi's wisdom, has illuminated the critical shift in cybersecurity from a reactive stance to one of proactive engagement and assertive action. Like a skilled warrior who dictates the terms of engagement rather than merely parrying blows, modern cybersecurity demands that we take the fight to the enemy.

We began by exploring **proactive defense**, recognizing that anticipating and neutralizing threats before they materialize is paramount. Penetration testing and red teaming serve as our controlled simulations, stress-testing our defenses and revealing vulnerabilities before malicious actors can exploit them. The crucial follow-through of vulnerability remediation and patch management ensures that discovered weaknesses are addressed swiftly and efficiently. Furthermore, security automation and orchestration act as our force multipliers, enabling rapid and precise responses to the ever-increasing volume and velocity of cyber threats.

Moving beyond mere prevention, we delved into **offensive security**, emphasizing the critical importance of understanding the attacker mindset. By comprehending their motivations, tactics, and objectives, we gain a significant advantage in anticipating their moves and crafting more effective defenses. Threat intelligence and hunting further empower us to proactively seek out lurking threats and disrupt their operations before they can inflict harm. The ethical considerations of ethical hacking and responsible disclosure underscore our commitment to using offensive techniques for defensive good.

Finally, we highlighted the indispensable role of **collaboration and information sharing** in amplifying our disruption capabilities. In a landscape where attackers thrive in the shadows and exploit the lack of coordination, our collective defense, built upon trust and the open exchange of threat intelligence, becomes an unbreakable chain, transforming isolated efforts into a unified force capable of dismantling even the most sophisticated operations.

Just as Musashi emphasized the importance of understanding fire – its power, its unpredictability, and the necessity of controlling it – so too must we understand the "fire" of cyberattacks. This chapter has underscored that true mastery in cybersecurity, like the mastery of swordsmanship, lies not just in reacting to the flames but in proactively shaping the environment, anticipating the enemy's moves, and taking decisive action to extinguish threats before they can spread. By embracing the proactive and assertive strategies outlined in this "Fire Book," we move beyond the role of mere defenders and become active participants in shaping a more secure digital future.

Key 4: Threats (The Wind Book)

"If you know the enemy and know yourself, you need not fear the result of a hundred battles."
— Sun Tzu

The wind, in its gentler moods, whispers secrets and carries the scent of distant rain. It can be a playful companion, rustling leaves in a soothing melody. But the wind also has a darker side, a raw power that can tear through landscapes, leaving devastation in its wake. In this untamed aspect of the wind, we find the essence of our fourth key: Threats.

For Musashi, the concept of threat was not an abstract notion confined to dusty scrolls or philosophical debates. It was a tangible force, as real as the steel of an opponent's blade or the gnawing hunger in his belly during lean times. He had faced threats in myriad forms – the sudden ambush in a moonlit forest, the subtle manipulation of a cunning rival, the crushing weight of expectation from those who believed in his potential.

Just as the wind can shift without warning, danger can also emerge from unexpected quarters. In an instant, what appears calm and predictable can transform into a tempest of adversity. This key, Threats, will delve into the nature of these forces that seek to undermine our progress, challenge our resolve, and ultimately, test the very foundations of who we are. Through Musashi's experiences and the lessons we can glean from the unpredictable nature of the wind, we will explore how to recognize, understand, and ultimately navigate the inevitable threats that life throws our way. Prepare yourself, for the winds are rising, and the true test of resilience is about to begin.

The Evolving Threat Landscape

Current Trends and Future Predictions

Just as the winds never cease to shift and change direction, so too does the landscape of threats that we face. A new, unforeseen challenge today might overshadow what was considered a significant danger yesterday. Understanding these evolving trends and attempting to predict what lies ahead is crucial for navigating the complexities of our world, whether we are a lone warrior like Musashi facing down an opponent or a global society grappling with multifaceted risks.

Here, we will explore some of the key current trends in the threat landscape and offer some predictions for the future.

Current Trends

The Digital Storm: Cyber threats continue to proliferate and become more sophisticated. We are seeing a rise in ransomware attacks targeting critical infrastructure, supply chain vulnerabilities being exploited, and the increasing use of artificial intelligence to craft more convincing phishing schemes and malware. Nation-state actors are also becoming more adept at using cyber operations for espionage, disruption, and even sabotage. This digital storm is constantly brewing, with new vulnerabilities being discovered and exploited at an alarming rate.

Environmental Turbulence: The impacts of climate change are no longer a distant threat but a present reality. Extreme weather events like hurricanes, wildfires, and droughts are becoming more frequent and intense, causing widespread devastation and displacement. Resource scarcity, particularly water and arable land, is also exacerbating tensions and creating new vulnerabilities. This environmental turbulence is a slow but relentless force, reshaping our world.

Geopolitical Crosswinds: Global power dynamics are shifting, leading to increased competition and instability. Existing conflicts are becoming more protracted and complex, and new areas of tension are emerging. The rise of nationalism and protectionism is disrupting international cooperation and creating further uncertainty. This geopolitical crosswind can bring sudden and unpredictable shifts in the global landscape.

The Information Whirlwind: The rapid spread of misinformation and disinformation through online platforms poses a significant threat to social cohesion and democratic processes. Malicious actors are increasingly adept at manipulating public opinion and sowing discord. This information whirlwind can quickly erode trust and create societal fragmentation.

Biological Shifting Sands: The COVID-19 pandemic highlighted the ever-present threat of infectious diseases. The potential for new pandemics or the resurgence of existing ones remains a significant concern. Furthermore, advancements in biotechnology also raise concerns about the potential for misuse, creating new biological threats. These biological shifting sands remind us of the fragility of our health and well-being.

Future Predictions

Hyper-Personalized Cyberattacks: We can expect to see AI playing an even larger role in cyberattacks, leading to highly personalized and sophisticated threats that are harder to detect and defend against. Attacks targeting individuals' specific vulnerabilities and exploiting their online behavior will likely become more common.

Climate Cascade: The interconnectedness of environmental threats will become more apparent. For example, droughts leading to food shortages, which in turn could fuel social unrest and migration. We will likely see more "climate cascades" where one environmental crisis triggers a series of others.

The Weaponization of Information Deepens: The ability to create realistic fake videos and audio (deepfakes) will likely be further refined, making it even harder to distinguish between truth and falsehood. This could be used to manipulate elections, damage reputations, and incite violence on an unprecedented scale.

Increased Geopolitical Fragmentation: The trend towards multipolarity and competition between major powers is likely to continue, potentially leading to a more fragmented and less stable global order. This could increase the risk of regional conflicts and make international cooperation on global issues more challenging.

The Convergence of Threats: We may see different types of threats converging and amplifying each other. For instance, a cyberattack targeting critical infrastructure during a natural disaster could have catastrophic consequences. Understanding these interconnections will be crucial for effective threat mitigation.

Just as a skilled navigator learns to read the subtle shifts in the wind to anticipate storms, we must constantly monitor and analyze the evolving threat landscape. By understanding the current trends and thoughtfully considering future possibilities, we can better prepare ourselves and build resilience against the inevitable challenges that lie ahead. The winds of threat are ever-present, and our ability to adapt and respond will determine our survival and progress.

The Rise of Ransomware and Malware

The digital threat landscape is increasingly dominated by the insidious and financially motivated ransomware attacks and the pervasive damage caused by a wide array of malware. The rise of these malicious software types represents a significant escalation in cyber threats, impacting not just individuals but also crippling businesses, disrupting critical infrastructure, and posing a severe risk to national security. Ransomware, in particular, has evolved from a relatively minor concern to a multi-billion-dollar criminal industry, characterized by increasingly sophisticated tactics, larger ransom demands, and a growing focus on high-value targets. This malware operates by encrypting a victim's data, effectively holding it hostage until a ransom is paid, typically in an untraceable cryptocurrency. The pressure to pay is immense, as organizations face the prospect of prolonged operational downtime, significant financial and reputational losses, and the potential exposure of sensitive information.

The scope of ransomware attacks has broadened dramatically, moving beyond individual users to target organizations of all sizes and sectors aggressively. According to analysis from cybersecurity firms like Black Kite the frequency and cost of ransomware incidents have seen a concerning upward trend. Their research often highlights the interconnectedness of the modern business ecosystem, emphasizing how vulnerabilities in one organization's security posture can be exploited to launch attacks on its partners and suppliers, leading to widespread disruption across entire supply chains. The focus on third-party risk is particularly relevant in ransomware, as attackers frequently target smaller, less secure vendors as a gateway to larger, more lucrative targets. The financial implications extend far beyond the ransom payment itself, encompassing recovery costs, legal fees, reputational damage, and regulatory fines, often dwarfing the initial demand.

Beyond the immediate financial impact, ransomware attacks can cause significant operational disruption, potentially leading to the shutdown of essential services. The targeting of critical infrastructure, such as healthcare, energy, and transportation, has become a major concern for governments, security agencies, companies, and individuals worldwide. As Black Kite's analysis often points out, the cascading effects of a successful ransomware attack on a critical service provider can have far-reaching and potentially life-threatening consequences. Furthermore, the rise of "Ransomware-as-a-Service" (RaaS) models has led to "professionalization" of ransomware operations and lowered the barrier to entry for cybercriminals, allowing less technically skilled individuals to launch sophisticated attacks using pre-built tools and infrastructure provided by more experienced operators. This has fueled the rapid proliferation of ransomware and made it a more pervasive and persistent threat.

While ransomware focuses on data encryption and extortion, the broader category of malware encompasses various malicious software designed to compromise computer systems in various ways. This includes viruses that spread by attaching to legitimate files, worms that can self-replicate and propagate across networks, Trojan horses that disguise themselves as harmless software to gain access, and spyware that secretly monitors user activity and steals sensitive information. The sophistication of malware continues to evolve, with attackers employing advanced techniques to evade detection by traditional antivirus software and security tools. Polymorphic malware can change its code to avoid signature-based detection, while fileless malware operates in memory, leaving fewer traces on the hard drive. The increasing use of artificial intelligence and machine learning by cybercriminals is also leading to the development of more adaptive and evasive malware strains. The combined threat of ransomware and increasingly sophisticated malware necessitates a multi-layered security approach, encompassing proactive prevention measures, robust detection capabilities, and effective incident response plans to mitigate the risks and minimize the potential damage from these ever-evolving digital adversaries.

Cloud-based threats and vulnerabilities

The rapid adoption of cloud computing has revolutionized the way organizations store, process, and access data and applications. While offering numerous benefits such as scalability, flexibility, and cost-effectiveness, this shift has also introduced a new set of threats and vulnerabilities that organizations must address. The very nature of cloud environments, with their shared infrastructure and reliance on internet connectivity, creates unique security challenges that differ significantly from traditional on-premises models. Understanding these specific risks is crucial for organizations looking to leverage the power of the cloud securely.

One of the primary areas of concern in cloud security is the **misconfiguration of cloud services**. The complexity and vast array of options available in cloud platforms can easily lead to errors in configuration, such as leaving storage buckets publicly accessible, mismanaging identity and access controls, or failing to properly secure network configurations. These misconfigurations can create significant vulnerabilities that attackers can exploit to gain unauthorized access to sensitive data or launch further attacks. Unlike on-premises environments where security configurations are often more centralized and controlled, the distributed and dynamic nature of the cloud requires a deep understanding of the specific security responsibilities shared between the cloud provider and the customer. Furthermore, the **shared**

responsibility model in cloud computing can sometimes lead to confusion and gaps in security. While cloud providers are responsible for the security of the underlying infrastructure, customers are typically responsible for securing their data, applications, and configurations within the cloud environment. A lack of clarity about these shared responsibilities can result in critical security controls being overlooked. Organizations need to clearly define their security responsibilities and ensure they have the necessary expertise and tools to secure their cloud deployments effectively.

Another significant category of cloud-based threats revolves around **data breaches and data loss.** With large volumes of sensitive data being stored in the cloud, these environments become prime targets for cybercriminals. Data breaches can occur due to various factors, including misconfigurations, vulnerabilities in cloud services, compromised credentials, or insider threats. Additionally, data loss can occur due to accidental deletion, service outages, or even the termination of a cloud subscription without proper data migration. Ensuring the confidentiality, integrity, and availability of data in the cloud requires robust encryption mechanisms, strong access controls, and comprehensive data backup and recovery strategies.

Compromised credentials and account takeovers are also prevalent threats in cloud environments. Many organizations rely on username and password combinations for authentication, which can be susceptible to phishing attacks, brute-force attempts, or data breaches on other platforms. Once an attacker gains access to legitimate cloud credentials, they can potentially access sensitive data, launch malicious activities, or even take control of entire cloud accounts. Implementing multi-factor authentication (MFA) and adopting strong identity and access management (IAM) policies are crucial steps in mitigating this risk.

Finally, **insider threats and third-party risks** also pose significant concerns in the cloud. Employees with malicious intent or accidental errors can lead to data breaches or security incidents. Similarly, organizations that rely on third-party applications or services integrated with their cloud environment need to ensure that these external entities also adhere to strong security standards. Vulnerabilities in third-party integrations can be exploited to gain access to the organization's cloud resources. Addressing these cloud-specific threats and vulnerabilities requires a proactive and comprehensive security strategy that encompasses robust access controls, continuous monitoring, regular security assessments, and a clear understanding of the shared responsibility model.

The Human Element Revisited

Social Engineering Tactics and Defense

While technological advancements have introduced many sophisticated cyber threats, the human element remains a critical vulnerability that attackers frequently exploit through social engineering tactics. These techniques manipulate individuals into performing actions or divulging confidential information that can compromise security. Unlike purely technical attacks that target software or hardware vulnerabilities, social engineering preys on human psychology, leveraging trust, fear, urgency, or helpfulness to achieve malicious goals. Understanding these tactics and implementing effective defenses is paramount in building a robust security posture.

One of the most prevalent social engineering tactics is **phishing**, which involves sending deceptive emails, messages, or phone calls that appear to be from legitimate sources, such as banks, online retailers, or even colleagues. These messages often aim to trick recipients into clicking malicious links, opening infected attachments, or providing sensitive information like passwords or credit card details. A more targeted form of phishing, known as **spear phishing**, focuses on specific individuals or groups within an organization, using personalized information to increase the likelihood of success. **Whaling** is a variant that targets high-profile individuals, such as executives, who often have privileged access to sensitive data.

Another common tactic is **pretexting,** where an attacker creates a believable scenario or pretext to convince a victim to provide information or access. This might involve impersonating a colleague, a help desk technician, or a representative from a trusted organization. The attacker often researches their target to gather information that makes their pretext more convincing. **Baiting** involves offering something enticing, such as a free download or a USB drive found in a public place, containing malware. Curiosity or the desire for a free item can lead victims to compromise their systems unknowingly.

Tailgating, also known as piggybacking, is a physical social engineering tactic where an unauthorized individual follows an authorized person into a restricted area. This often relies on politeness or a reluctance to question someone who appears to belong. Finally, **quid pro quo** involves offering a benefit in exchange for information or access. For example, an attacker might pose as a survey taker offering a small reward in exchange for personal details.

Defending against social engineering attacks requires a multi-faceted approach focusing on technical controls and human awareness. **Security awareness training** is crucial for educating employees about common social engineering tactics, recognizing red flags, and verifying requests for sensitive information. Regular training and simulated phishing exercises can help reinforce these lessons and build a security-conscious culture within an organization.

Implementing **strong policies and procedures** can also help mitigate the risk of social engineering. This includes policies on password management, data handling, and verifying unusual requests. Requiring multiple forms of verification for sensitive transactions or information requests can make it harder for attackers to succeed. Technical controls, such as spam filters, anti-phishing software, and multi-factor authentication, can also help block or detect social engineering attempts. However, technology alone is not sufficient, as attackers are constantly developing new and more sophisticated tactics. Ultimately, fostering a **security-conscious culture** where individuals are empowered to question suspicious requests and report potential threats is the most effective defense against the ever-evolving landscape of social engineering attacks.

Insider Threats and Mitigation

Insider threats represent a significant and often underestimated danger to organizations. Unlike external attacks originating from outside the network perimeter, insider threats stem from individuals who have legitimate access to an organization's systems, data, and physical locations. This privileged access makes insider threats particularly challenging to detect and prevent, as these individuals often know where sensitive information resides and how to bypass security controls. Insider threats can be broadly categorized into three main types: malicious insiders, negligent insiders, and compromised insiders.

Malicious insiders intentionally cause harm to the organization. This could involve stealing confidential data for personal gain or to sell to competitors, sabotaging systems, or disrupting operations out of spite or for ideological reasons. **Negligent insiders**, on the other hand, unintentionally create security vulnerabilities through carelessness, lack of awareness, or failure to follow security policies. This might include clicking on suspicious links, sharing passwords, or leaving sensitive data unprotected. Finally, **compromised insiders** are legitimate users whose accounts or devices have been taken over by external attackers. These attackers then leverage the insider's access to carry out their malicious activities.

Detecting insider threats can be difficult because the activities of malicious or compromised insiders may initially appear to be normal user behavior. However, there are certain indicators that organizations can look for. These might include unusual access patterns, such as accessing files or systems outside of normal working hours or job responsibilities, attempts to download or transfer large amounts of data, unauthorized software installations, or repeated violations of security policies. Behavioral analytics and user and entity behavior analytics (UEBA) tools can help identify these anomalies by establishing baseline behavior and flagging deviations that could indicate a potential insider threat.

Mitigating the risk of insider threats requires a comprehensive and multi-layered approach that addresses people, processes, and technology. Strong access controls are fundamental, ensuring that employees only have access to the systems and data they absolutely need to perform their job duties based on the principle of least privilege. Regular reviews and updates of access privileges are also essential, especially when employees change roles or leave the organization. Continuous monitoring of user activity, network traffic, and data movements is crucial for detecting suspicious behavior. This can involve using security information and event management (SIEM) systems, data loss prevention (DLP) tools, and user activity monitoring software.

Data loss prevention (DLP) tools can help prevent sensitive information from leaving the organization's control, whether intentionally or unintentionally. These tools can identify and block unauthorized attempts to copy, transfer, or share confidential data. Implementing **thorough background checks** on new hires can help reduce the risk of hiring malicious individuals. However, it's important to remember that an employee's circumstances and motivations can change over time, so ongoing monitoring and security awareness training are still necessary.

Employee training and awareness programs play a vital role in mitigating both malicious and negligent insider threats. Educating employees about security policies, the risks associated with insider threats, and how to report suspicious activity can help create a security-conscious culture. Emphasizing the importance of data protection and responsible use of company resources can also reduce the likelihood of unintentional security breaches. Finally, having well-defined **incident response plans** specifically for insider threats is crucial for containing and remediating any incidents that do occur in a timely and effective manner. This includes procedures for investigating potential insider threats, securing compromised systems, and recovering any lost or damaged data.

The Psychology of Cyberattacks: Beyond Bits and Bytes

To truly understand the persistent and evolving threat of cyberattacks, we must delve beyond the technical intricacies of malware and network vulnerabilities and confront the underlying psychology that drives both the perpetrators and the targets. Cyberattacks are not simply about code and algorithms; they are fundamentally human endeavors, fueled by a complex interplay of motivations, perceptions, and vulnerabilities. Examining this psychological landscape offers a deeper insight into why these attacks are so prevalent and how we can better defend against them.

On the offensive side, the motivation behind launching cyberattacks is as varied as the individuals and groups involved. For some, the primary driver is **financial gain** through ransomware extortion, financial data theft, or illicit cryptocurrency mining. This pursuit of profit can be a powerful motivator, often coupled with a sense of impunity afforded by the internet's relative anonymity and global reach. Others are driven by ideological or **political motivation**s, seeking to disrupt or damage adversaries, spread propaganda, or make a statement through hacktivism. The digital realm provides a platform for these actors to project their beliefs and exert influence on a scale previously unimaginable. Still others may be motivated by **revenge**, targeting former employers or perceived wrongdoers, while some individuals are simply drawn to the thrill and challenge of breaching security systems, viewing it as a test of their skills and intelligence. The anonymity offered by the digital world can also contribute to a sense of **deindividuation**, where individuals feel less constrained by social norms and are more likely to engage in risky or unethical behavior they might otherwise avoid in the physical world.

Conversely, understanding the **psychology of the victims** is crucial for comprehending the effectiveness of many cyberattacks, particularly those involving social engineering. Attackers often exploit fundamental human tendencies such as **trust,** our inherent inclination to believe and cooperate with others. Phishing emails, for example, often impersonate trusted entities to trick recipients into divulging sensitive information. **Urgency** is another powerful psychological lever, with attackers creating a sense of immediate threat or opportunity to bypass critical thinking and encourage impulsive actions. The **appeal to authority**, where individuals are more likely to comply with requests from perceived figures of power, is also frequently exploited. Furthermore, the **fear of missing out (FOMO)** can be weaponized, as seen in scams promising exclusive deals or access to limited resources. Even simple curiosity can lead individuals to click on malicious links or open infected attachments.

The psychological impact of being a victim of a cyberattack can be significant. Individuals may experience **fear, anxiety, and a loss of trust** in online services and technologies. Organizations can suffer **reputational damage** and financial distress following a data breach or ransomware attack. Understanding these psychological consequences can inform the development of more empathetic and effective communication strategies for dealing with victims. Moreover, by recognizing the psychological vulnerabilities that attackers exploit, we can design more effective security awareness training programs that empower individuals to recognize and resist social engineering tactics. Ultimately, acknowledging the profound psychological dimension of cyberattacks allows us to move beyond a purely technical understanding of the threat and develop more holistic and human-centric security strategies.

Specific Threats and How to Counter Them

Here, we will break down some of the most prevalent and emerging threats in the cybersecurity landscape and outline concrete steps organizations and individuals can take to mitigate their risk. This is a high-level review as there are significant resources available to learn more, some of which I have included in Appendix E.

Phishing and Email Security

The Threat: Phishing remains one of the most effective and widely used attack vectors. It involves deceptive communications, often via email, designed to trick recipients into divulging sensitive information, clicking malicious links, or opening infected attachments. Different forms of phishing include:

- General Phishing: Broadly distributed emails targeting a large number of recipients.

- Spear Phishing: Highly targeted emails aimed at specific individuals or groups, often leveraging personalized information.

- Whaling: A type of spear phishing that targets high-level executives or individuals with significant access and authority.

- Smishing (SMS Phishing): Phishing attacks conducted via text messages.

- Vishing (Voice Phishing): Phishing attacks carried out over the phone.

How to Counter Them

Technical Controls

- **Robust Email Filtering:** Implement advanced email security solutions that can identify and block known phishing attempts, spam, and malicious attachments. This includes Sender Policy Framework (SPF), DomainKeys Identified Mail (DKIM), and Domain-based Message Authentication, Reporting & Conformance (DMARC) to verify sender authenticity

- **Link Analysis and Sandboxing**: Employ technologies that analyze links within emails for malicious content before users click them and sandbox attachments in isolated environments to detect malware.

- **Anti-Malware and Anti-Virus Software**: Ensure all endpoints have up-to-date anti-malware and anti-virus software to detect and remove any malicious software that may bypass initial email filtering.

- **Multi-Factor Authentication (MFA):** Implement MFA on email accounts to add an extra layer of security, making it harder for attackers to gain access even if they obtain passwords.

- **Email Encryption:** Use encryption for sensitive communications to protect the content of emails in transit and at rest.

Human-Focused Measures

- **Security Awareness Training:** Conduct regular and engaging training sessions to educate users about different types of phishing attacks, red flags to look for (e.g., suspicious sender addresses, grammatical errors, urgent requests), and how to report suspicious emails.
- **Phishing Simulation Exercises:** Regularly conduct simulated phishing campaigns to test employee awareness and identify areas where further training is needed.
- **Promote a Culture of Skepticism:** Encourage users to be cautious about unsolicited emails and to independently verify requests for sensitive information through official channels.
- **Establish Clear Reporting Mechanisms:** Make it easy for employees to report suspicious emails or messages to the IT security team for investigation.
- **Implement Clear Policies and Procedures:** Define clear policies regarding email usage, handling sensitive information, and reporting security incidents.

Web Application Attacks and Security

- **The Threat**: Web applications are a frequent target for attackers due to their direct exposure to the internet and the often-sensitive data they handle. Common web application attack vectors include:

    o **SQL Injection (SQLi):** Exploiting vulnerabilities in database queries to gain unauthorized access to or manipulate data.

    o **Cross-Site Scripting (XSS):** Injecting malicious scripts into websites viewed by other users.

    o **Cross-Site Request Forgery (CSRF):** Tricking users into performing unintended actions on a web application they are authenticated to.

- **Broken Authentication and Session Management:** Flaws in how users are authenticated and sessions are managed, allowing attackers to impersonate legitimate users.
- **Security Misconfiguration:** Incorrectly configured servers, applications, or cloud services that create vulnerabilities.
- **Insecure Deserialization:** Exploiting vulnerabilities in how applications handle serialized data.
- **Using Components with Known Vulnerabilities**: Utilizing outdated or vulnerable libraries and frameworks.
- **Insufficient Logging and Monitoring:** Lack of adequate logging and monitoring makes it difficult to detect and respond to attacks.

How to Counter Them

Secure Development Practices

- **Security by Design:** Integrate security considerations throughout the entire software development lifecycle (SDLC).
- **Code Reviews**: Conduct thorough code reviews to identify and address potential security vulnerabilities before deployment.
- **Input Validation and Sanitization:** Implement strict input validation and sanitization to prevent malicious data from being processed by the application.
- **Parameterized Queries or Prepared Statements**: Use parameterized queries or prepared statements to prevent SQL injection attacks.
- **Output Encoding:** Encode output properly to prevent XSS attacks.
- **Secure Session Management:** Implement strong session management mechanisms, including using secure session IDs, setting appropriate timeouts, and regenerating session IDs after authentication.

Security Testing and Assessment

- **Static Application Security Testing (SAST):** Analyze source code for potential vulnerabilities without executing the application.
- **Dynamic Application Security Testing (DAST):** Test the running application for vulnerabilities from an attacker's perspective.
- **Penetration Testing:** Employ ethical hackers to simulate real-world attacks and identify weaknesses in the application's security.
- **Vulnerability Scanning:** Regularly scan web applications and their underlying infrastructure for known vulnerabilities.

Deployment and Configuration Security

- **Secure Configuration:** Properly configure web servers, application servers, and databases according to security best practices.
- **Patch Management**: Regularly patch and update all software components, including operating systems, web servers, application frameworks, and libraries.
- **Web Application Firewall (WAF):** Deploy a WAF to filter malicious traffic and protect against common web application attacks.
- **Rate Limiting and Throttling:** Implement rate limiting and throttling to prevent brute-force attacks and denial-of-service attempts.

Monitoring and Logging

- **Comprehensive Logging:** Implement detailed logging of all application activity, including user actions, errors, and security events.
- **Real-time Monitoring and Alerting:** Monitor logs and application performance for suspicious activity and set up alerts for potential security incidents.

Data Breaches and Data Loss Prevention

- **The Threat:** Data breaches involve unauthorized access, disclosure, or theft of sensitive information. Data loss can occur for various reasons, including cyberattacks, accidental deletion, insider threats, or natural disasters. The consequences of data breaches and data loss can be severe, including financial losses, reputational damage, legal liabilities, and regulatory fines.

How to Counter Them

Data Identification and Classification

- **Identify Sensitive Data:** Determine what data is considered sensitive and requires protection (e.g., personal identifiable information (PII), financial data, intellectual property).
- **Data Classification:** Categorize data based on its sensitivity level to apply appropriate security controls.

Access Control and Authorization

- **Principle of Least Privilege:** Grant users only the minimum level of access necessary to perform their job duties.
- **Role-Based Access Control (RBAC):** Assign access permissions based on roles rather than individual users, simplifying management.
- **Strong Authentication and Authorization Mechanisms:** Implement strong passwords, MFA, and robust authorization controls to verify user identities and enforce access restrictions.
- **Regular Access Reviews:** Periodically review and revoke unnecessary access privileges.

Data Encryption

- **Encryption at Rest:** Encrypt sensitive data stored on servers, databases, and storage devices.
- **Encryption in Transit:** Encrypt data transmitted over networks, including internal and external communications (e.g., using TLS/SSL).

Data Loss Prevention (DLP) Solutions

- **Implement DLP Tools:** Deploy DLP software to monitor and control the flow of sensitive data, preventing unauthorized copying, sharing, or transmission.
- **Endpoint DLP:** Monitor and control data on user workstations and devices.
- **Network DLP:** Monitor and control data traversing the network.
- **Cloud DLP:** Monitor and control data stored in and accessed through cloud services.

Data Backup and Recovery

- **Regular Backups:** Implement a comprehensive backup strategy to regularly back up critical data.
- **Secure Backup Storage:** Store backups in a secure and separate location, ideally offline or in a geographically diverse location.
- **Regular Restore Testing:** Periodically test the data restoration process to ensure its effectiveness.

Incident Response Planning

- **Develop an Incident Response Plan:** Create a detailed plan outlining the steps to be taken in the event of a data breach or data loss incident.
- **Regularly Test and Update the Plan:** Conduct tabletop exercises and simulations to test the plan and update it based on lessons learned.

Physical Security

- **Secure Physical Access:** Control physical access to data centers and other sensitive areas.
- **Proper Disposal of Media:** Implement secure procedures for the disposal of old hard drives, tapes, and other media containing sensitive data.

Emerging Threats (AI-Powered Attacks, Quantum Computing)

- **The Threat:** The rapid advancements in artificial intelligence (AI) and quantum computing are poised to introduce new and potentially more potent threats to cybersecurity.

AI-Powered Attacks

- **AI-Enhanced Phishing:** AI can be used to create more convincing and personalized phishing emails and social engineering attacks.
- **Automated Malware Generation:** AI could automate the creation of new and evasive malware variants.
- **AI-Driven Network Attacks:** AI algorithms could be used to identify vulnerabilities and launch sophisticated attacks on network infrastructure.
- **Evasion of AI-Based Defenses:** Attackers could use adversarial AI techniques to craft attacks that can bypass AI-powered security systems.
- **Deepfakes for Social Engineering:** AI-generated realistic fake videos and audio could be used to impersonate individuals and manipulate victims.

Quantum Computing

Quantum computers will break today's encryption (like RSA/ECC), making stored sensitive data vulnerable later ("Harvest Now, Decrypt Later"). New quantum-resistant standards (PQC) are ready. Migrating is complex and takes years.

Action needed now: Plan your PQC transition (inventory, risk assessment).

- **Breaking Current Encryption Algorithms:** Quantum computers have the potential to break many of the currently used public-key cryptography algorithms, such as RSA and ECC, which are fundamental to securing online communications and data. This could render vast amounts of currently encrypted data vulnerable.

- **New Cryptographic Attacks:** Quantum algorithms could be developed to launch novel types of cryptographic attacks.

How to Counter Them

AI-Powered Defenses

- **Leverage AI for Threat Detection:** Employ AI and machine learning algorithms to analyze network traffic, user behavior, and other data to identify and respond to sophisticated attacks.
- **AI-Driven Anomaly Detection:** Use AI to establish baseline behavior and detect deviations that may indicate malicious activity.
- **Automated Threat Response:** Utilize AI to automate certain aspects of incident response, such as isolating infected systems or blocking malicious traffic.
- **Develop Robust AI Security:** Implement security measures to protect AI systems themselves from adversarial attacks and data poisoning.

Preparing for Quantum Computing Threats

- **Research and Monitor Quantum Computing Advancements:** Stay informed about the progress in quantum computing and its potential implications for cybersecurity.
- **Invest in Post-Quantum Cryptography (PQC):** Begin exploring and implementing post-quantum cryptographic algorithms that are believed to be resistant to attacks from quantum computers. NIST (National Institute of Standards and Technology) is currently working on standardizing PQC algorithms.
- **Hybrid Approaches:** Consider adopting hybrid cryptographic approaches that combine classical and post-quantum algorithms to provide a transition path.
- **Long-Term Data Protection Strategies:** For highly sensitive data with long retention requirements, consider the potential future threat from quantum computing and plan accordingly.
- **Collaboration and Information Sharing:** Engage with industry peers, research institutions, and government agencies to stay

ahead of emerging quantum threats and best practices for mitigation.

Addressing these specific and emerging threats requires a proactive, layered, and adaptive security strategy that combines technical controls, robust processes, and a strong security-aware culture. Continuous learning and adaptation are crucial in the ever-evolving landscape of cybersecurity.

Third-Party Risk Management (TPRM)

Organizations increasingly rely on a vast network of third-party vendors for various services, technologies, and data processing. This reliance, while offering numerous benefits in terms of efficiency and specialization, also introduces significant risks that must be carefully managed through a robust Third-Party Risk Management (TPRM) program. The importance of TPRM cannot be overstated, as breaches and vulnerabilities within a third-party's environment can have a cascading effect, directly impacting the security, operations, reputation, and regulatory compliance of the primary organization. A weakness in a vendor's cybersecurity posture can serve as an open door for attackers to gain access to sensitive data or disrupt critical services, highlighting the necessity of extending an organization's security perimeter to encompass its entire third-party ecosystem.

Traditional approaches to TPRM often relied heavily on qualitative assessments, such as questionnaires and subjective evaluations of a vendor's security controls. While these methods provide a foundational understanding of a vendor's security practices, they often fall short in providing a comprehensive and quantifiable view of the actual cyber risk. Moving beyond these qualitative assessments is crucial for organizations to make informed decisions about vendor selection, ongoing monitoring, and risk mitigation strategies. A shift towards more quantitative and data-driven approaches allows for a more objective understanding of the likelihood and potential impact of cyber threats originating from third parties. This involves leveraging security ratings, threat intelligence feeds, and other quantifiable metrics to gain a more accurate picture of a vendor's security posture.

To enhance the agility and responsiveness of TPRM programs, organizations can look to frameworks like the OODA Loop (Observe, Orient, Decide, Act). Originally developed for military strategy, the OODA Loop provides a cyclical approach to decision-making that can be effectively applied to the dynamic nature of third-party risks. By continuously observing the evolving threat landscape and a vendor's security posture, orienting oneself to the relevant risks and context, deciding on appropriate actions, and then acting on those decisions, organizations can create a more adaptive and proactive TPRM process. This agile approach allows for quicker identification of emerging threats and vulnerabilities within the third-party ecosystem and facilitates timely responses to mitigate potential risks before they escalate into incidents.

A critical component of moving towards a more quantitative approach in TPRM is the adoption of Cyber Risk Quantification (CRQ). CRQ aims to assign financial values to cyber risks, enabling organizations to understand the potential economic impact of a third-party breach. By quantifying these risks in monetary terms, organizations can prioritize their TPRM efforts based on the potential financial exposure, justify investments in security controls, and make more informed decisions about risk acceptance or mitigation. CRQ provides a common language for discussing cyber risk with business stakeholders and allows for a more strategic allocation of resources to address the most significant threats within the third-party ecosystem.

The integration of Artificial Intelligence (AI) and Machine Learning (ML) technologies is further revolutionizing the field of TPRM. AI and ML can be leveraged to automate various aspects of the TPRM process, such as the initial assessment of a large number of vendors, the continuous monitoring of their security posture through real-time threat intelligence analysis, and the identification of anomalies or potential risks that might be missed by manual processes. These technologies can analyze vast amounts of data from various sources to provide a more comprehensive and up-to-date view of third-party risks, enabling organizations to scale their TPRM efforts effectively and proactively identify potential threats before they are exploited.

Ultimately, building a robust TPRM framework requires a holistic and integrated approach that encompasses policy, process, technology, and people. This framework should begin with clearly defined policies and procedures for vendor onboarding, risk assessment, ongoing monitoring, and offboarding. It should incorporate a mix of qualitative and quantitative assessment methods, leveraging tools and technologies for automation and continuous monitoring.

Furthermore, a robust program requires the involvement and collaboration of various stakeholders across the organization, including procurement, legal, IT, and security teams. A successful TPRM framework is not a one-time implementation but rather an ongoing process of continuous improvement, adaptation, and vigilance to effectively manage the ever-evolving risks associated with third-party relationships.

Key 4: Threats - A Summary Through the Eyes of Musashi

Key 4, Threats, much like a sudden, violent gust of wind, explores the myriad dangers that can buffet and break us. Just as the wind can shift from a gentle breeze to a destructive tempest without warning, so too can threats emerge from unexpected quarters, testing our resilience and resolve. For Musashi, a life lived on the edge of a blade was a constant negotiation with threat in its most immediate form. Yet, as this key reveals, the landscape of danger is far more intricate than a simple clash of steel.

The evolving threat landscape mirrors the unpredictable nature of the wind itself. What posed a danger yesterday might be superseded by a new, unforeseen peril today. Musashi – navigating treacherous landscapes and ever-shifting political climates – would have understood this implicitly. The rise of ransomware and malware, like an unseen, digital squall, can cripple and extort, leaving devastation in its wake. For

Musashi, this might have manifested as an unexpected ambush or a poisoned well, an insidious attack that strikes at the core of one's ability to survive. Similarly, cloud-based threats – vulnerabilities lurking within seemingly vast and powerful systems – could be likened to hidden weaknesses in a seemingly impenetrable fortress, a concept Musashi would have grasped in his strategic engagements.

The human element, revisited within this key, underscores that the most potent threats often wear a human face. Social engineering tactics, the art of manipulation and deception, are akin to the cunning words of a rival seeking to undermine Musashi's honor or exploit his trust. Just as he had to discern friend from foe, so too must we learn to recognize and defend against these psychological attacks. Insider threats, dangers lurking within one's own ranks, resonate with the theme of betrayal, a bitter wind that can shatter even the strongest alliances, a reality Musashi likely faced in his complex relationships.

Understanding the psychology of cyberattacks offers a deeper insight into the motivations and vulnerabilities at play. Musashi, a keen observer of human nature, would have sought to understand his opponents' desires, fears, and weaknesses to gain an advantage. This same principle applies to the digital realm, where recognizing the psychological levers that attackers pull is crucial for effective defense. Finally, Third-Party Risk Management, while seemingly a modern concept, echoes the challenges Musashi would have faced in choosing his allies and understanding the potential dangers associated with those he relied upon. Just as a weak link in a chain can compromise the whole, a vulnerability in a third-party vendor can expose an entire organization to risk.

In essence, Key 4: Threats, through the lens of Musashi's world in The Wind Book, reveals that danger is not a static entity but a dynamic and multifaceted force. Whether it manifests as a physical confrontation, a subtle manipulation, or a digital assault, the ability to recognize, understand, and ultimately navigate these threats is paramount. Like a seasoned warrior learning to read the currents of the wind to anticipate storms, we must cultivate a keen awareness of the threat landscape to protect ourselves and our endeavors in an increasingly complex and interconnected world. The wind, in its turbulent nature, serves as a constant reminder that vigilance and preparedness are our most steadfast defenses against the inevitable storms of life.

Key 5: Process (The Book of the Void)

> *"Study strategy over the years and achieve the spirit of the warrior. Today is victory over yourself of yesterday; tomorrow is your victory over lesser men." — Miyamoto Musashi*

Entering the Flow of the Void

Having journeyed through the preceding keys, we now arrive at a crucial juncture: **Process**. In the context of this exploration of the Void, "Process" might initially seem at odds with the very notion of emptiness and formlessness that Musashi so eloquently describes in his final scroll. Yet, it is within the understanding of process that we begin to grasp how the abstract principles of the Void manifest in the tangible world.

Musashi's "Book of the Void" isn't simply about nothingness; it's about a state of being unburdened by preconceptions, allowing for effortless action and adaptation. But how does one achieve this state, and more importantly, how does one operate from it? The answer we propose lies in understanding the underlying processes that govern our actions, our learning, and our very interaction with reality.

This chapter will go into the dynamic nature of process as it relates to the Void. We will explore how a deep understanding of flow, adaptation, and continuous refinement can lead us closer to the freedom and clarity that Musashi envisioned. Just as a swordsman must master the fundamental movements before transcending them into intuitive action, so too must we understand the processes that shape our experiences before we can truly embody the emptiness of the Void.

Prepare to examine the cyclical nature of learning, the importance of iterative action, and the subtle art of aligning our processes with the boundless potential of the Void. For it is through the conscious engagement with process that we move from merely contemplating the Void to actively living within its liberating embrace.

Building a Security Culture

Creating a truly effective security culture transcends the mere implementation of security tools and policies; it necessitates a fundamental shift in mindset and behavior across the entire organization. This transformation is driven by a multi-faceted approach, with **leadership and communication** forming the bedrock upon which a strong security culture is built. Effective leaders at all levels must actively champion security, demonstrating through their actions and words that it is a top priority. This means more than just signing off on security budgets; it requires actively participating in security initiatives, consistently reinforcing security messaging, and integrating targeted security considerations into all business decisions.

Communication must be transparent, frequent, and tailored to different audiences within the organization. Executive leadership should articulate the strategic importance of security and its alignment with overall business objectives. Middle management plays a crucial role in translating these high-level messages into actionable guidance for their teams, ensuring that security practices are integrated into daily workflows. Open channels of communication must be established to encourage employees to report suspicious activities or voice security concerns without fear of reprisal. This fosters a culture of psychological safety where everyone feels empowered to contribute to the collective security posture. Regular updates on the threat landscape, security incidents (while respecting confidentiality), and the effectiveness of security measures can help maintain awareness and build trust. Leading by example is paramount; if leaders consistently adhere to security protocols, it sets a powerful precedent for the rest of the organization.

Building upon this foundation, comprehensive **training and awareness** programs are essential for equipping employees with the knowledge and skills necessary to navigate the complex landscape of cyber threats. These programs should move beyond generic, infrequent presentations and instead adopt a more engaging and continuous approach. Training should be tailored to specific roles and responsibilities within the organization, recognizing that different departments and individuals face unique security risks. For instance, the training needs of the finance department, handling sensitive financial data, will differ from those of the marketing team. A variety of training methods should be employed, including interactive online modules, engaging in-person workshops, simulated phishing exercises, and practical hands-on labs. Awareness campaigns should be creative and memorable, utilizing a mix of communication channels such as posters, intranet banners, videos, and even gamified challenges to keep security top-of-mind. Regular reminders and updates on emerging threats and best practices are crucial to reinforce learning and prevent complacency. The effectiveness of training and awareness programs should be continuously evaluated through metrics such as employee participation rates, performance in simulated phishing exercises, and the number of reported security incidents. Feedback from employees should be actively sought to ensure the programs remain relevant and impactful. The goal is to cultivate a security-conscious workforce where individuals understand their role in protecting organizational assets and are empowered to make secure decisions in their daily work.

Finally, a robust security culture integrates **incentives and accountability** to encourage desired security behaviors and address non-compliance. Incentives can take various forms, from public recognition for individuals or teams who demonstrate exemplary security practices to tangible rewards such as bonuses or gift cards. Gamified challenges that reward secure behaviors, such as identifying and reporting simulated phishing emails, can also be highly effective in fostering engagement and friendly competition. The key is to design incentive programs that genuinely motivate employees and reinforce the desired security outcomes. Alongside incentives, clear lines of accountability are essential. Organizations must establish well-defined security policies and procedures with clear consequences for violations. Accountability should be applied fairly and consistently across all levels of the organization. This includes holding individuals responsible for their actions, whether intentional or negligent, that compromise security. However, accountability should not be solely punitive; it should also be viewed as an opportunity for learning and improvement. When security incidents occur, a blame-free post-incident review process can help identify root causes and prevent future occurrences. It's crucial to strike a balance between holding individuals accountable and fostering a supportive environment where mistakes are seen as learning opportunities rather than grounds for immediate punishment.

This balanced approach encourages a sense of shared responsibility for security and promotes a culture where everyone is invested in protecting the organization's assets. By thoughtfully implementing incentives and accountability measures, organizations can solidify their security culture and create a workforce that is both motivated and responsible in maintaining a secure environment.

Governance and Compliance

Security Policies and Procedures

A truly resilient and trustworthy organization understands that security is not an afterthought but an intrinsic element of its very foundation, meticulously woven into the fabric of its operations through robust governance and compliance frameworks. Governance, in this context, encompasses the overarching system of rules, practices, and processes by which an organization directs and controls its security activities. It establishes the roles, responsibilities, and decision-making authority related to information security, ensuring that security objectives are aligned with broader business goals and that risks are effectively managed at an organizational level. This includes defining the strategic direction for security, allocating resources appropriately, and overseeing the implementation and effectiveness of security controls. Compliance, on the other hand, focuses on adhering to a multitude of external and internal requirements, including legal and regulatory mandates, industry standards, contractual obligations, and internal policies. Demonstrating compliance is not merely a matter of ticking boxes; it signifies a commitment to operating ethically and responsibly, minimizing legal and financial risks, and maintaining the trust of stakeholders, including customers, partners, and regulatory bodies.

The landscape of governance and compliance is constantly evolving, driven by emerging threats, technological advancements, and changes in legal and regulatory frameworks. Therefore, organizations must adopt a proactive and adaptive approach, continuously reviewing and updating their governance and compliance mechanisms to remain effective and relevant. Strong leadership is paramount in fostering a culture of governance and compliance, setting the tone from the top and ensuring that security is viewed as everyone's responsibility. Regular audits, both internal and external, play a critical role in assessing the effectiveness of governance and compliance frameworks, identifying areas for improvement, and providing assurance to stakeholders.

At the heart of any effective governance and compliance framework lie well-defined security policies and procedures, serving as the tangible articulation of an organization's security commitments and operational guidelines.

Security policies are formal, high-level statements that articulate an organization's stance on specific security topics. They establish the fundamental principles, objectives, and expectations for behavior related to the protection of information assets. These policies are typically approved by senior management, signifying their organizational importance and mandatory nature. Effective security policies are characterized by their clarity, conciseness, and enforceability. They should be written in plain language, avoiding technical jargon where possible, and should clearly state what is required or prohibited. Furthermore, they should be regularly reviewed and updated to reflect changes in the threat landscape, business operations, and regulatory requirements. [1]

A comprehensive set of security policies will address a wide range of security domains, including but not limited to: data security (covering classification, handling, and disposal), network security (addressing access control, firewall management, and intrusion detection), physical security (outlining measures for protecting physical assets and facilities), personnel security (addressing employee onboarding, offboarding, and background checks), incident response (defining the process for handling security breaches), and business continuity and disaster recovery (outlining plans for maintaining operations in the event of disruptions). These policies provide the overarching framework for security decision-making and set boundaries for acceptable behavior within the organization.

While security policies define the 'what' and 'why' of security requirements, security procedures provide the detailed 'how-to' instructions for implementing and adhering to those policies. They are step-by-step guides that outline the specific actions that individuals must take to comply with security policies and perform security-related tasks effectively. Well-written security procedures are accurate, complete, user-friendly, and readily accessible to those who need to follow them. They should provide enough detail to ensure consistency in execution while remaining practical and easy to understand. Examples of security procedures directly corresponding to the policy examples mentioned earlier could include: a procedure for creating and managing strong passwords, a procedure for encrypting sensitive data at rest and in transit, a procedure for configuring firewall rules, a procedure for granting and revoking physical access to facilities, a step-by-step guide for reporting a suspected security incident, and detailed instructions for performing data backups and system recovery.

The development of security procedures involves input from subject matter experts and those performing the tasks to ensure practicality and accuracy. Regular review and updates are crucial to maintain relevance as technology and business processes evolve. Effective communication and training are essential to ensure that personnel know the relevant security procedures and understand how to follow them correctly.

Security procedures serve as a critical mechanism for ensuring the consistent application of security controls across the organization, reducing the risk of errors and omissions, and ultimately contributing to a stronger and more resilient security posture. By clearly defining both the principles (through policies) and the practical steps (through procedures), organizations can establish a solid foundation for effective governance and compliance, leading to enhanced security and greater trust among stakeholders.

Audits and Assessments: Unleashing the Inner Security Detective (With More Gadgets!)

Alright, buckle up, security enthusiasts! We're jump deeper into the thrilling world of audits and assessments – think of it as upgrading our security detective from a lone wolf with a magnifying glass to a whole team equipped with high-tech gadgets and forensic skills! Remember how we said audits are like top-to-bottom investigations? Let's unpack that a bit more.

An audit typically follows a structured process. First comes the planning phase, where the scope and objectives of the audit are defined. What are we trying to check? What rules are we making sure we're following? Then comes the fieldwork, where the actual investigation happens. Auditors will examine documentation, interview staff, and perform tests to see if the security controls work as intended. For example, they might check whether access to sensitive data is appropriately restricted to authorized personnel or if the process for making changes to critical systems is secure and well-documented. They'll be looking for things like proper audit trails (like a digital breadcrumb trail of who did what and when) and ensuring that security logs are properly maintained and reviewed. After the fieldwork, the auditors compile their findings into a report, highlighting any weaknesses or areas for improvement. Finally, there's the follow-up, where the organization takes action to address the identified issues, and the auditors may verify that those actions have been effective.

Both internal audits, conducted by your own team, and external audits, performed by independent experts, bring valuable perspectives. Internal audits offer continuous monitoring and a deep understanding of the organization, while external audits provide an unbiased and objective evaluation, often required for compliance with specific regulations. The frequency and scope of audits depend on various factors, including the size and complexity of the organization, the sensitivity of the data it handles, and the applicable regulatory requirements.

Now, let's crank up the excitement with assessments! If audits are a comprehensive check-up, assessments are specialized diagnostic tests. We talked about penetration testing, where ethical hackers try to find and exploit vulnerabilities, think of it as a controlled break-in to see how strong your defenses really are! There are different flavors of pen testing: black box (the testers have no prior knowledge of your systems, like a real attacker), white box (they have full knowledge, allowing for a more targeted and efficient assessment), and grey box (a mix of both). The process usually involves several stages: reconnaissance (gathering information about the target), scanning (identifying potential entry points), exploitation (trying to actually break in), post-exploitation (seeing what they can do once inside), and finally, a detailed report outlining their findings and recommendations.

Vulnerability scanning is another key assessment tool. Imagine this process as using an automated scanner to quickly identify known weaknesses in your software and systems. There are active scans that probe systems and passive scans that analyze network traffic. These scans can uncover a wide range of vulnerabilities, from outdated software to misconfigurations. Risk assessments are like our strategic security planning sessions. We identify our valuable assets, the potential threats they face, and the vulnerabilities that could be exploited. Then, we calculate the level of risk associated with each threat and prioritize our efforts to mitigate the most critical ones. Security architecture reviews are like getting an expert to look at the blueprint of our security systems to ensure they are well-designed and implemented effectively. These reviews examine how different security components work together and identify any potential flaws in the overall design. And for a truly holistic view don't forget social engineering assessments, which test the human element of security by simulating attacks like phishing emails or phone scams to see how susceptible employees are.

The key to a successful assessment is clearly defining the scope and objectives. What exactly are we trying to evaluate? What are our goals for this assessment? Once the assessment is complete, the findings are typically prioritized based on their severity and the potential impact they could have on the organization. Then comes the crucial step of remediation, where we take action to fix the identified vulnerabilities and weaknesses. Like a detective solving a case, we need to follow through and ensure the problems are resolved! Various tools and methodologies are used in assessments, from specialized software for vulnerability scanning and penetration testing to established frameworks for risk management.

So, while audits and assessments might sound a bit like homework, they are incredibly valuable tools that help us stay ahead of the curve in the ever-evolving world of cyber threats. They provide us with the insights we need to strengthen our defenses, improve our security posture, and ultimately sleep a little better at night knowing our digital world is a bit safer, thanks to our dedicated security detectives and their awesome gadgets!

Regulatory Requirements (GDPR, HIPAA, etc.): Navigating the Legal Labyrinth of Security

In today's interconnected world, securing information isn't just a matter of best practice; it's often a legal imperative. Organizations worldwide are subject to a complex web of regulatory requirements designed to protect sensitive data, ensure privacy, and maintain the integrity of various industries. These aren't just suggestions; they are laws, rules, and guidelines imposed by governmental bodies and industry organizations, and failing to comply can result in significant penalties, legal repercussions, and irreparable damage to reputation.

Think of regulations like GDPR (General Data Protection Regulation), which primarily focuses on the data protection and privacy of individuals within the European Union. It dictates how personal data must be collected, processed, stored, and secured, granting individuals significant rights over their information. Similarly, in the United States, HIPAA (Health Insurance Portability and Accountability Act) sets the standard for protecting sensitive patient health information. It outlines specific security and privacy rules that healthcare providers, insurance companies, and other covered entities must adhere to. Beyond these widely known examples, numerous other regulations exist, such as PCI DSS (Payment Card Industry Data Security Standard) for organizations that handle credit card information, various state-specific data breach notification laws, and industry-specific regulations depending on the sector (e.g., financial services, energy).

The importance of these regulatory requirements cannot be overstated. They are in place to protect individuals from harm, ensure the responsible handling of sensitive information, and foster trust in organizations that collect and process data. Compliance with these regulations demonstrates a commitment to ethical practices and a respect for the rights of individuals. Conversely, non-compliance can lead to hefty fines, costly legal battles, and significant reputational damage, potentially eroding customer trust and impacting the bottom line.

For any organization, the **first crucial step is to identify which regulatory requirements apply to its specific operations, industry, and geographic locations**. This requires a thorough understanding of the legal and regulatory landscape. Once the applicable regulations are identified, organizations must then implement appropriate security controls, policies, and procedures to meet the specific requirements outlined in those regulations. This often involves a multi-faceted approach, encompassing technical safeguards, administrative controls, and physical security measures.

It's vital to recognize that regulatory compliance is not a one-time event but an ongoing process. Regulations evolve, new threats emerge, and business operations change. Therefore, organizations must continuously monitor their compliance status, adapt their security practices as needed, and stay informed about any updates or changes to the relevant regulations. As we discussed earlier, regular audits and assessments play a crucial role in verifying compliance and identifying potential gaps.

Understanding and adhering to regulatory requirements is a fundamental aspect of a comprehensive security strategy. It's not just about avoiding penalties; it's about building a culture of responsibility, protecting sensitive information, and maintaining the trust of customers, partners, and the wider community. Ignoring these legal obligations is not an option in today's environment; instead, organizations must embrace them as essential pillars of their overall security and governance framework.

The Future of Cybersecurity: Navigating a Shifting Landscape

Cybersecurity is constantly in flux, driven by relentless innovation on both the offensive and defensive fronts. Understanding the trajectory of this evolution is crucial for individuals, organizations, and even nations to safeguard their digital assets and maintain operational resilience. Let's explore the key aspects shaping the future of this critical domain.

Emerging Trends and Technologies

The threat landscape is becoming increasingly sophisticated and multifaceted. We are witnessing a surge in advanced persistent threats (APTs) orchestrated by state-sponsored actors and highly organized cybercriminal groups, often employing novel techniques and targeting critical infrastructure. Ransomware continues to evolve, with threat actors employing double and even triple extortion tactics, targeting data encryption, exfiltration, and denial-of-service attacks. The interconnectedness of our world through supply chains has also emerged as a significant vulnerability, where a compromise at one point in the chain can have cascading effects across numerous organizations. The proliferation of Internet of Things (IoT) devices, often with weak security protocols, presents a vast and expanding attack surface. Furthermore, the rise of deepfakes poses a new challenge, blurring the lines between reality and fabrication, potentially leading to sophisticated social engineering attacks and disinformation campaigns.

On the technological horizon, several emerging trends will profoundly impact cybersecurity. Quantum computing – while still in its nascent stages – holds the potential to break current encryption algorithms, necessitating the development and adoption of quantum-resistant cryptography. Advancements in network security are leading to the rise of zero-trust architectures, which operate on the principle of "never trust, always verify," and the adoption of Secure Access Service Edge (SASE) frameworks, converging network security and wide area network (WAN) capabilities into a cloud-delivered service. The increasing convergence of IT (Information Technology) and OT (Operational Technology) in industrial control systems presents unique security challenges as traditionally isolated OT environments become more connected to the internet. Decentralized technologies like blockchain offer potential security benefits through their inherent immutability and transparency, but also introduce new complexities and potential attack vectors. The burgeoning metaverse will create new digital environments and economies, inevitably attracting cybercriminals and requiring novel security approaches for virtual identities, assets, and interactions. Finally, privacy-enhancing technologies (PETs) are gaining traction, offering innovative ways to analyze and utilize data while preserving individual privacy. This will be crucial in navigating increasingly stringent data protection regulations.

The Role of AI and Automation

Artificial intelligence (AI) and automation are poised to revolutionize cybersecurity, offering both immense opportunities for enhancing defenses and potential avenues for more sophisticated attacks. On the defensive side, AI is being leveraged for advanced threat detection and analysis. Machine learning algorithms can analyze vast amounts of data to identify anomalies, patterns of malicious behavior, and indicators of compromise that might be missed by traditional security tools. Behavioral analysis, powered by AI, can establish baselines of normal activity and flag deviations that could indicate a security breach. Automated incident response and remediation are becoming increasingly critical in dealing with the speed and scale of modern cyberattacks. AI-powered systems can automatically isolate infected devices, block malicious traffic, and even initiate rollback procedures, significantly reducing the time it takes to contain and recover from an incident. AI is also enhancing threat intelligence by automatically aggregating and analyzing data from various sources to provide a more comprehensive and timely understanding of emerging threats and attacker tactics. Furthermore, automation streamlines repetitive security tasks, such as patching vulnerabilities, managing configurations, and responding to common security alerts, freeing human security analysts to focus on more complex and strategic issues.

However, the rise of AI also presents a double-edged sword. Cybercriminals are increasingly exploring the use of AI for offensive purposes. AI-powered malware could potentially evade traditional detection mechanisms by learning and adapting its behavior. Deepfake technology could be weaponized to create highly convincing phishing attacks, impersonating trusted individuals or organizations. The ability of AI to generate realistic text and images could also be used to spread disinformation and manipulate public opinion. This necessitates a constant arms race, with cybersecurity professionals needing to develop and deploy AI-powered defenses that can effectively counter AI-driven attacks. Ethical considerations surrounding the use of AI in cybersecurity are also paramount. Issues such as bias in AI algorithms, the potential for autonomous decision-making in security incidents, and the need for human oversight must be carefully addressed to ensure responsible and effective deployment of these technologies.

Preparing for the Unknown

In the face of such a dynamic and unpredictable landscape, the most crucial aspect of future cybersecurity is preparing for the unknown. This requires a fundamental shift in mindset from reactive defense to proactive resilience. Organizations need to build security architectures that are inherently adaptable and can withstand unforeseen threats. This includes embracing principles like security by design and defense in depth, ensuring that security is integrated into every stage of development and that multiple layers of security controls are in place. Continuous learning and skill development for cybersecurity professionals are paramount. The threat landscape is constantly evolving, and security teams must stay abreast of the latest trends, technologies, and attack techniques. Achieving this requires investment in training, certifications, and fostering a culture of continuous improvement.

Proactive threat intelligence and anticipation will be key to staying ahead of the curve. Organizations need to actively monitor the threat landscape, analyze emerging trends, and anticipate potential future attacks. This involves leveraging threat intelligence feeds, participating in information sharing communities, and even conducting their own threat hunting activities. Collaboration and information sharing across industries and between public and private sectors will be crucial in collectively preparing for future threats. Sharing threat intelligence, best practices, and lessons learned can help the entire cybersecurity community become more resilient. Robust incident response plans that are regularly tested and updated are essential for effectively managing and recovering from security incidents, regardless of their nature or origin. These plans need to be flexible and adaptable to handle novel and unexpected attacks.

Ultimately, fostering a security-conscious culture across the entire organization is a critical element of preparing for the unknown. Every employee needs to understand their role in maintaining security and be empowered to identify and report potential threats. This requires ongoing awareness training, clear communication, and a culture where security is seen as everyone's responsibility. Finally, it's important to acknowledge the potential impact of geopolitical factors and future global events on cybersecurity. International tensions, cyber warfare, and global crises can all have significant implications for the threat landscape, requiring organizations to be prepared for a wide range of potential scenarios. By embracing adaptability, continuous learning, collaboration, and a proactive security posture, we can better navigate the uncertainties of the future and build a more secure digital world.

Key 5: Process - A Summary Through the Eyes of Musashi

Thinking about Key 5: Process, concerning Musashi's Book of the Void, it strikes me that the 'Process' here isn't just about following a checklist of security tasks. Instead, it feels much more like the fundamental journey of truly grasping the essence of security, almost like striving for that state of unobstructed understanding Musashi describes as the Void.

When considering leadership and communication through this lens, the process isn't merely about giving orders or dictating policies. It's about guiding the team towards a shared understanding of security's importance, communicating the core principles so clearly that they resonate and become part of everyone's mindset.

Training and awareness become more than just mandatory sessions. They represent the continuous, dedicated practice that builds an almost instinctual understanding of secure practices. It's akin to Musashi's relentless dedication to his sword techniques, where repetition leads to effortless mastery. Awareness, in this context, is that constant state of readiness, that intuitive ability to recognize potential threats.

Incentives and accountability shift from being just external motivators to reflecting a more profound, personal commitment to security. The real reward is the enhanced strength and resilience achieved through diligent practice, and accountability becomes about taking ownership of our role in maintaining that security, driving continuous improvement from within.

Governance and compliance, while they provide the necessary framework and structure, are the initial forms we engage with to build a foundational understanding. The true aim is to internalize the underlying principles of order and security so thoroughly that our actions become less about strict adherence to rules and more about an intuitive application of those principles.

Audits and assessments take on the role of critical self-reflection, a process of emptying our minds to objectively evaluate our defenses. This effort rigorously tests our understanding and identifies areas where we can refine our approach, allowing for constant striving for a clearer, more accurate perception of our security posture.

Even regulatory requirements, which might initially seem like external pressures, become part of the broader landscape we must skillfully navigate. Mastering the processes to meet these requirements contributes to our overall understanding of the security environment and allows us to focus on the more fundamental aspects of protection.

And when we look at the future of cybersecurity, the connection to the Void becomes even more apparent. The threat landscape is constantly shifting, new technologies emerge, and the only truly effective way to prepare is to cultivate that state of openness and adaptability that Musashi describes. Continuous learning, experimentation, and refinement become the constant, allowing us to meet the unknown with a flexible and receptive mindset.

Ultimately, when viewed through the lens of Musashi's Book of the Void, this Key of Process isn't just about the steps we take in cybersecurity. It's about the ongoing journey towards a deeper, more intuitive understanding of the art of defense, constantly striving for that state of clarity and unobstructed perception that allows us to act effectively and decisively in the face of any challenge.

Mastering the Way

"The road to mastery is traveled not in leaps, but in practiced   steps   taken   with   intent."
— Bob Maley

The culmination of our exploration through these five keys – Self-Knowledge (The Ground Book), Agility (The Water Book), Action (The Fire Book), Threats (The Wind Book), and Process (The Book of the Void) – mark not an end, but a significant milestone on a far greater journey. Like the aspiring samurai diligently studying the terrain (Ground), adapting like water, striking with the force of fire, being aware of the wind's whispers, and ultimately seeking the emptiness of true understanding, we have laid the groundwork for a deeper mastery of cybersecurity. But as Musashi – a lifelong seeker of perfection – would attest, the true Way is one of perpetual motion, an unceasing quest for excellence. To genuinely master the digital battlefield and ascend to the level of a true cyber strategist demands an unwavering commitment to principles that resonate deeply with the spirit of his teachings.

The very essence of Musashi's philosophy begins with understanding the ground beneath one's feet, and so too must our journey in cybersecurity commence with **Self-Knowledge** (The Ground Book). This foundational key emphasizes the critical importance of knowing oneself. In our context, this means a deep and comprehensive understanding of our organization, systems, data, and inherent vulnerabilities. Just as a warrior must know his strengths and limitations, the cyber strategist must possess a clear and honest assessment of their digital terrain. This involves meticulous inventory, thorough documentation, and constantly evaluating our security posture. Without this solid foundation of self-knowledge, any subsequent actions will be built on shifting sands that are vulnerable to unseen weaknesses. This pursuit of understanding our digital landscape is not a one-time endeavor, but an ongoing process of discovery and refinement.

Building upon this foundation is the principle of **Agility** (The Water Book). Musashi likened strategy to water, emphasizing its ability to adapt to any container, to flow around obstacles, and to erode even the strongest defenses powerfully. In the dynamic realm of cybersecurity, agility is paramount. The threat landscape constantly shifts, with new attacks emerging and old vulnerabilities being exploited in novel ways. The cyber strategist must be like water, capable of adapting their defenses, responding quickly to incidents, and adjusting their strategies as circumstances dictate. This requires flexible architectures, adaptable policies, and a mindset that embraces change rather than resisting it. Agility also speaks to the importance of communication and collaboration, allowing teams to coordinate effectively and respond swiftly to evolving threats.

With a firm understanding of ourselves and the ability to adapt, we move to **Action** (The Fire Book). Musashi understood that knowledge and adaptability are meaningless without decisive action. Like fire, which is both powerful and swift, the cyber strategist must be capable of executing their plans with purpose and clarity. This involves the timely implementation of security controls, the decisive response to security incidents, and the proactive pursuit of threat mitigation. Action in cybersecurity requires not only technical proficiency but also courage and the ability to make difficult decisions under pressure. It is about translating knowledge and agility into tangible results, effectively defending against attacks and protecting valuable assets. However, just as fire must be controlled and directed, so too must our actions in cybersecurity be guided by strategy and a clear understanding of the potential consequences.

Before action, however, comes awareness of the environment, embodied by **Threats** (The Wind Book). Musashi spoke of the importance of perceiving the unseen, of understanding the subtle movements and intentions of the opponent. In cybersecurity, this translates to a deep awareness of the threat landscape. The cyber strategist must be attuned to the whispers of the digital wind, constantly monitoring for emerging threats, analyzing attack trends, and understanding the motivations and tactics of adversaries. This requires robust threat intelligence capabilities, proactive vulnerability scanning, and a keen understanding of the human element as a potential attack vector. Like the wind revealing hidden dangers, a thorough understanding of threats allows us to anticipate attacks, strengthen our defenses in vulnerable areas, and make informed decisions about our security posture.

Finally, all these elements coalesce into **Process** (The Book of the Void). As Musashi taught, the Void is not emptiness in the sense of nothingness, but a state of being open, adaptable, and free from fixed ideas. In cybersecurity, Process represents the continuous journey of learning, refinement, and applying the other four keys in a holistic and integrated manner. It is the understanding that security is not a static state but an ongoing cycle of assessment, adaptation, action, and awareness. The cyber strategist who embraces the Void understands that mastery is not about adhering to rigid rules but about developing an intuitive understanding of the underlying principles, allowing them to act decisively and effectively in any situation. This requires a commitment to continuous improvement, a willingness to embrace challenges, and the cultivation of a strategic mindset that sees beyond immediate forms and understands the interconnectedness of all things in the digital realm.

Mastering the Way of cybersecurity, therefore, is a lifelong pursuit, a continuous cycle of building Self-Knowledge, cultivating Agility, taking decisive Action, understanding Threats, and striving for the encompassing understanding of Process. The five keys we have explored provide a comprehensive framework, mirroring the profound wisdom of Musashi. Embrace this journey with diligence, courage, and an unyielding thirst for understanding, and you will walk the path of a true cyber strategist.

The Unseen Edge: Blending the OODA Loop, the Book of Five Rings, and the Five Keys of Cyber Strategy

"Machines don't fight wars. Terrain does not fight wars. Humans fight wars. You must get into the mind of humans. That's where the battles are won." — Col. John Boyd

I would be remiss if I didn't discuss the OODA Loop in the context of the Five Rings. My goal was to pick up and finish what I started in 2010; however, the rigid layout of five sequential keys may lead one to believe it is a fixed loop, which is not the case in this book, nor in the use of the OODA Loop.

To truly master the art of cyber strategy, one must move beyond a rigid interpretation of the OODA Loop. Colonel John Boyd's concept is not a simple, sequential checklist; it's a dynamic and iterative cycle, a continuous interplay of observation, orientation, decision, and action. Understanding its fluidity and the critical role of cognitive factors is paramount.

In the high-stakes arena of cyber warfare, where milliseconds can determine victory or defeat, a purely technical approach is often insufficient. The unseen edge, the decisive advantage, lies in the ability to think strategically, to anticipate the adversary's moves, and to operate within their decision cycle with superior speed and understanding. This mastery is achieved by weaving together the pragmatic dynamism of the OODA Loop with the profound wisdom of Miyamoto Musashi's Book of Five Rings and the focused insights of our Five Keys of Cyber Strategy: Self-Knowledge (The Ground Book), Agility (The Water Book), Action (The Fire Book), Threats (The Wind Book), and Process (The Book of the Void). This powerful synthesis creates a holistic framework for cyber strategy, enabling practitioners to not only react to threats but also shape the digital battlefield and proactively achieve true dominance.

The OODA Loop's power lies in its ability to generate speed and agility, creating a crucial temporal advantage over the adversary. It's about disrupting their tempo, getting inside their decision cycle to sow confusion, uncertainty, and ultimately, disorder. This requires a deep understanding of the cognitive aspects of conflict, a domain where Musashi's emphasis on perceiving the opponent's mind, exploiting their weaknesses, and maintaining one's own mental clarity finds a potent modern application. Just as Musashi sought to anticipate his opponent's intentions before their sword moved, the cyber strategist aims to disrupt the adversary's OODA Loop, rendering their actions predictable, slow, and ultimately, ineffective or irrelevant.

Orientation: The Decisive Element of Comprehensive Understanding

Boyd astutely identified Orientation as the most critical phase of the OODA Loop. It transcends mere situational awareness, encompassing the complex interplay of our genetic predispositions, cultural influences, and accumulated prior experiences. In the context of cybersecurity, this translates to a multifaceted understanding that extends beyond the technical intricacies of threats. It demands a deep comprehension of the adversaries' motivations – whether they are driven by financial gain, geopolitical espionage, ideological beliefs, or simple disruption. It requires insight into their thought processes, their likely strategies, and their potential reactions to our defensive maneuvers. This crucial phase resonates profoundly with Musashi's emphasis on knowing oneself and the enemy.

Our Key 1: Self-Knowledge (The Ground Book) forms the bedrock of our Orientation. This involves a thorough understanding of our network topology, our critical data flows, the sensitivity of our assets, the security protocols in place, and the behavioral patterns of our users. It necessitates honest assessments of our strengths and weaknesses, our security gaps, and the cultural norms within our organization that might either bolster or hinder our security posture. Simultaneously, understanding the adversaries' motivations, their likely tools and techniques, their preferred attack vectors, and their overarching strategic goals aligns perfectly with the user's Key 4: Threats (The Wind Book), where we diligently seek to perceive the unseen forces acting against us, much like sensing the subtle shifts in the wind. This requires continuous threat intelligence gathering, analysis of attacker TTPs (Tactics, Techniques, and Procedures), and an understanding of the broader geopolitical and cybercrime landscape.

Furthermore, Boyd's inclusion of genetic heritage, cultural influences, and prior experiences underscores the vital role of historical context in cybersecurity. Our organizational history of security incidents, our ingrained security practices (both effective and ineffective), and the broader cultural attitudes towards risk all shape our current Orientation. Understanding past attack trends, threat actors' evolution, and previous defensive strategies' successes and failures provides invaluable context for interpreting current observations and anticipating future threats. This historical lens, combined with a granular understanding of our environment and the multifaceted nature of the threats we face, allows us to build a richer, more nuanced, and ultimately more accurate mental model of the digital battlefield.

Cognitive Warfare and Deception: Shaping the Narrative of the Digital Conflict

Cyber conflict is increasingly a cognitive battle, a contest for perception and understanding. It's about shaping the adversary's view of reality, manipulating the information they receive, and exploiting inherent vulnerabilities in human decision-making processes. Deception, therefore, becomes a powerful tool in the cyber strategist's arsenal. By creating ambiguity, misdirection, and false narratives, we can disrupt an adversary's Orientation, leading to delayed, flawed, or entirely inappropriate decisions. This strategic use of deception echoes Musashi's tactical brilliance in creating openings through feints, diversions, and the calculated exploitation of his opponent's expectations. Just as Musashi would use subtle shifts in his stance, feigned attacks, and psychological pressure to unbalance his opponent and create an exploitable opening, the cyber strategist can employ a range of deception techniques to sow confusion and disrupt the adversary's OODA Loop.

This cognitive battlefield is where the user's Key 2: Agility (The Water Book) becomes particularly relevant. The ability to rapidly deploy and adapt deception tactics, and to react swiftly to the adversary's attempts at uncovering the truth, is crucial. By thoroughly understanding the adversary's Orientation – their assumptions about our defenses, their biases in interpreting data, and their expected courses of action – we can proactively shape the information they encounter, leading them down false paths, wasting their resources, and disrupting their decision-making process. This might involve the strategic deployment of honeypots that convincingly mimic valuable targets, the creation of decoy data designed to lure attackers away from critical assets, the use of false flags to misattribute attacks, or the carefully orchestrated misdirection of incident response teams to buy valuable time. However, the effective use of deception requires careful planning, a deep understanding of the adversary, and a keen awareness of the potential risks and ethical considerations involved.

The Fluidity of the Loop: Orchestrating the Tempo of Engagement

The OODA Loop's true power lies in its iterative and non-linear nature. It is not a rigid sequence of steps but a dynamic cycle where the phases of Observation, Orientation, Decision, and Action often occur simultaneously and interact in complex and unpredictable ways. A skilled cyber strategist understands this fluidity and can leverage it to their advantage, compressing their own decision cycle while deliberately stretching out the adversary's. This temporal mismatch creates a critical advantage, leading to the adversary's inability to effectively respond, ultimately resulting in their strategic or tactical collapse. This inherent fluidity mirrors Musashi's profound understanding of adapting to the flow of battle, much like water conforming to its container (The Water Book).

Compressing our own OODA Loop requires speed, decisiveness, and efficiency in Action (The Fire Book and Key 3). Once we have observed and oriented ourselves to a threat, the ability to make rapid and effective decisions and execute them with speed and precision is critical. This might involve the automated blocking of known malicious IP addresses, the swift isolation of infected systems based on pre-defined incident response plans, or the rapid deployment of security patches to address newly discovered vulnerabilities. Stretching the adversary's OODA Loop, on the other hand, leverages our Agility (The Water Book and Key 2) and our deep understanding of Threats (The Wind Book and Key 4) to introduce elements of confusion, uncertainty, and complexity into their perception of the environment. This might involve employing polymorphic malware that constantly changes its signature, using dynamic attack vectors that shift and evade detection, or strategically withholding or leaking information to create ambiguity about our defenses and intentions. Underpinning this entire dynamic interplay is the user's Key 5: Process (The Book of the Void). It is through a continuous cycle of observation, orientation, decision, and action, constantly learning from each iteration and adapting our strategies in real-time, that we truly master the rhythm of the digital battlefield and gain a decisive temporal advantage.

Developing Cognitive Agility: The Unseen Weapon of the Cyber Strategist

Cyber strategists must cultivate and continuously refine their cognitive agility to excel in this complex and rapidly evolving environment. This involves developing a specific set of mental capabilities that resonate strongly with Musashi's principles of mental discipline, clear judgment, and unwavering focus.

Critical Thinking: The ability to rigorously analyze information, identify subtle patterns amidst noise, and make sound judgments under intense pressure is paramount. This aligns directly with Musashi's emphasis on seeing the true nature of things, unclouded by emotion or preconceived notions. It also connects deeply with the user's Key 5: Process (The Book of the Void), where a mind free from distractions can perceive the underlying reality of a situation with greater accuracy. Cyber strategists must be able to dissect complex technical data, understand the implications of seemingly disparate events, and form well-reasoned conclusions to guide their actions.

Adaptability: The capacity to rapidly adjust strategies and tactics in response to new information, unexpected adversary actions, and changing circumstances is essential for survival and success. This mirrors Musashi's Water Book and the user's Key 2: Agility. Just as water flows and adapts to the terrain, the cyber strategist must be able to pivot their defenses, modify their plans, and embrace new approaches when faced with unforeseen challenges. This requires a flexible mindset and a willingness to abandon outdated strategies in favor of more effective ones.

Anticipation: A master strategist is skilled at thinking ahead, predicting potential adversary actions based on available intelligence and understanding their motivations, and proactively shaping the environment to our advantage. This resonates strongly with Musashi's Wind Book and the user's Key 4: Threats. By diligently studying our adversaries, understanding their likely objectives and capabilities, and analyzing their past behavior, we can anticipate their future moves and proactively implement countermeasures.

Understanding Human Factors: Recognizing the significant role of social engineering, human error, and inherent cognitive biases in most cybersecurity breaches is crucial for developing effective defenses. This connects to Musashi's awareness of his own and his opponent's spirit and the user's Key 1: Self-Knowledge (The Ground Book) as understanding our own human vulnerabilities and those of our organization is fundamental to building a robust security posture. Cyber strategists must understand how attackers exploit trust, manipulate emotions, and leverage human error to gain access to systems and data. This understanding informs the development of effective awareness training, the implementation of user-friendly security controls, and the creation of a strong security culture.

By embracing the OODA Loop's inherent fluidity and diligently mastering the cognitive aspects of conflict, all while being guided by the timeless wisdom of Musashi and the focused insights of the Five Keys of Cyber Strategy, cyber strategists can transcend mere reactive defense and achieve true mastery of the digital battlefield. This integrated and holistic approach provides an unseen edge, enabling proactive defense, effective deception, and ultimately, the ability to dictate the tempo and outcome of engagements in the constantly evolving and challenging cybersecurity landscape.

Final Thoughts and Words of Wisdom

"You can be somebody. Or you can do something."
— Col. John Boyd

The path of cybersecurity – like the Way of Strategy – unfolds through understanding and application. Remember the five keys that have guided us: Self-Knowledge (The Ground Book), the foundation upon which all else is built; Agility (The Water Book), the ability to adapt and flow with the changing tides; Action (The Fire Book), the decisive execution born of understanding; Threats (The Wind Book), the keen awareness of the forces that surround us; and finally, Process (The Book of the Void), the ultimate understanding that transcends form.

Let your journey begin with profound Self-Knowledge, understanding your own strengths and weaknesses, your systems and vulnerabilities. Cultivate Agility, like water adapting to its vessel, allowing you to respond effectively to any challenge. Embrace decisive Action, striking with purpose and clarity when the moment demands. Remain ever mindful of Threats, like the wind revealing all that is hidden. And finally, strive for the understanding of Process, that state of openness and adaptability embodied by the Void, where true mastery resides.

The digital landscape, like the battlefield, is ever in motion. Never stop learning, never become rigid in your approach, and consistently seek to understand the underlying principles governing this domain. For in the continuous application of these five keys, in the unwavering pursuit of understanding the Way, lies the path to becoming a true cyber strategist. Go forth, sharpen your awareness, and protect the digital realm with wisdom, agility, and resolute action.

Appendices

As mentioned previously, during the creation of this book, I utilized several AI Deep Research features to generate annotated content that informed my writing. I have included these research results here as unedited Appendices for the reader's review and to help understand how I arrived at some of my positions and conclusions.

Each appendix features extensive footnotes, and I have retained the annotation numbers within the appendix, rather than combining them as footnotes for the entire book.

This was a conscious decision on my part, as I wanted to keep each research article stand-alone, not only for my future use, but also for readers' use.

Links in the citations were active when the research was conducted, but may have changed or been modified beyond the author's control.

The size of the Appendices drives up the prices of the printed version, so I have also chosen to house those on the internet. They are available at https://www.c-ooda.com/book-of-five-keys. The Kindle and the hardcover version contain the full Appendices. The softcover contains the content of the Appendices without the pages of references and links.

As time allows, I will also be publishing the Appendices as articles on LinkedIn. You can check here -
https://www.linkedin.com/in/strategicciso/recent-activity/articles/

Appendix A: The Human Firewall in Cybersecurity – An Industry Analysis of its Validity and Relevance

1. Introduction: Defining the "Human Firewall"

The term "human firewall" in cybersecurity refers to the collective capability of an organization's employees to act as a primary line of defense against cyber threats.[1] This concept moves beyond the traditional reliance on technological safeguards, highlighting the crucial role each individual plays in maintaining a secure digital environment.[1] It encompasses the support and training provided to the workforce to ensure adherence to and active implementation of cybersecurity best practices, including the timely reporting of any suspicious cyber activity, whether potential threats originate internally or externally.[4] The term draws inspiration from the function of a traditional technological firewall, which serves as a barrier to prevent unauthorized access to a network or system; the human firewall, in parallel, acts as a human-based defense mechanism aimed at shielding against various cyber threats.[6]

The objective of this report is to conduct a comprehensive analysis of the "human firewall" concept within the realm of cybersecurity. This analysis will explore the historical emergence of the term, examine the perspectives of cybersecurity experts who advocate for and critique its validity, and ultimately assess its relevance in the context of contemporary cybersecurity practices. By delving into the origins, evolution, benefits, and limitations of the "human firewall," this report aims to provide cybersecurity professionals, IT managers, and business leaders with a thorough understanding of this concept and its implications for organizational security strategies.

2. The Historical Context and Evolution of the "Human Firewall" Concept

The term "human firewall" entered the cybersecurity lexicon with the growing understanding that security is not solely a technological concern but fundamentally involves people and the processes they follow.[7] Its origin lies in the increasing recognition that cybercriminals were shifting their focus towards exploiting the "human factor" through sophisticated social engineering tactics and direct human interaction[8] This realization underscored the point that even the most robust technical defenses could be circumvented if individuals within an organization were not vigilant and knowledgeable about potential threats.[9]

The development of the "human firewall" concept was driven by a significant shift in the cybersecurity landscape. Initially, organizational efforts to secure digital assets primarily revolved around the implementation of technical solutions, such as network firewalls and antivirus software, designed to ward off external threats.[10] However, as cyber threats evolved in complexity, threat actors began to increasingly target human vulnerabilities. Tactics like phishing and social engineering emerged as highly effective methods for gaining unauthorized access to sensitive information and systems, highlighting the critical need to educate and empower employees as a crucial first line of defense.[10] The rise of "people-centered attacks," which capitalize on innate human instincts such as curiosity and trust to trick users into clicking on malicious links, downloading harmful software, or divulging confidential information, further emphasized the necessity of a human-centric approach to security.[8] Moreover, the increasing prevalence of remote working arrangements has amplified the importance of a well-defined human firewall. With employees operating outside the traditional security perimeter, often relying on personal devices and less secure networks, individual awareness and adherence to security best practices have become paramount.[4]

The understanding and application of the "human firewall" have undergone considerable evolution over time. Initial approaches may have been limited to basic security awareness training sessions. However, the concept has matured into a more comprehensive and integrated strategy that emphasizes the need for continuous education, the cultivation of a strong security-conscious culture throughout the organization, and the empowerment of employees to proactively identify and report suspicious activities.[2] Contemporary strategies often incorporate principles of adult learning, change management methodologies, gamification techniques, and personalized training programs to foster lasting and effective security habits among employees.[3] The "human firewall" is now widely recognized as an indispensable component of any robust organizational cybersecurity strategy, considered a critical element that works in tandem with and complements traditional technological defenses.[2] This evolution reflects a deeper understanding of the human element in security and a move towards creating a more resilient and adaptable defense against an ever-changing threat landscape.

3. Arguments in Favor: The "Human Firewall" as a Vital Security Layer

Cybersecurity experts who champion the "human firewall" concept underscore that the human element frequently represents the most vulnerable point in an organization's security infrastructure, thereby making a well-trained and vigilant workforce an absolutely essential layer of defense.[2] In an era characterized by increasingly sophisticated, often AI-driven cyberattacks, these experts argue that human intuition, critical thinking capabilities, and the capacity for nuanced judgment remain indispensable assets that effectively complement the pattern recognition and data processing strengths of artificial intelligence in the detection and mitigation of threats.[11] Leading voices in the cybersecurity field advocate for a proactive approach that involves actively nurturing a security-aware culture within organizations and making sustained investments in continuous training initiatives designed to transform every employee into a frontline defender against cyber threats.[12]

Furthermore, the "human firewall" is viewed as a mechanism to standardize the involvement of all personnel in an organization's cyber defense efforts, thereby promoting a comprehensive and "human-first" approach to cybersecurity.[14]

The benefits of cultivating a security-aware workforce, often referred to as a "human firewall," are extensive and contribute significantly to an organization's overall security posture. Employees trained as part of the human firewall serve as the initial point of defense, equipped to recognize and appropriately respond to suspicious activities such as phishing attempts and social engineering tactics before these can infiltrate and compromise organizational systems and sensitive data.[2] This approach directly addresses the acknowledged "human factor," widely considered the weakest link in an organization's security. By fostering a culture of security awareness and shared responsibility, organizations can mitigate both technological and psychological vulnerabilities.[2] While technological security measures are undeniably vital, they are not infallible. A well-trained human firewall acts as a crucial supplementary layer, providing an additional level of defense against threats that may successfully bypass or exploit weaknesses in technical controls.[2] Moreover, employees who are part of a human firewall are better positioned to quickly identify and report potential security incidents, enabling the organization to mount a more swift and effective response, thereby minimizing the potential impact of a successful cyberattack.[2] Empowering all employees to actively participate in cybersecurity cultivates a security-conscious culture throughout the organization, fostering a collective sense of responsibility for safeguarding valuable assets, which is particularly critical in the face of increasingly elaborate and deceptive cyber threats.[2] Many industry-specific regulations and standards, such as HIPAA, PCI-DSS, and GDPR, mandate the implementation of comprehensive security awareness training programs for employees. Adopting a human firewall approach can assist organizations in meeting these crucial compliance requirements.[2] By equipping employees with the necessary knowledge and skills to recognize and avoid common cybersecurity threats, such as phishing emails and social engineering attacks, organizations can significantly mitigate the risk of human error leading to security breaches.[3] Unlike automated security tools that primarily detect known threats, a vigilant human firewall can identify novel or unique threats that might otherwise evade traditional detection methods.[3] Given the constantly evolving nature of the cyber threat landscape, ongoing training and awareness programs ensure that employees remain adaptable and informed about emerging threats.[3] In the event of a cybersecurity incident, a robust human firewall strategy can significantly enhance incident response efforts as prepared employees will be more likely to follow established protocols, report incidents promptly, and take appropriate action.[3] Furthermore, implementing a human firewall strategy can contribute to simplifying complex security protocols, making them more comprehensible and actionable for everyone within the organization.[14]

Human vigilance plays a pivotal role as both the initial point of defense and a crucial final safeguard against a wide spectrum of cyber threats. Employees often serve as the first to encounter various forms of attacks, including sophisticated phishing emails, subtle social engineering attempts designed to manipulate trust, and other suspicious activities that target human psychology.[2] In scenarios where advanced and persistent attacks manage to bypass an organization's technological defenses, a vigilant and well-informed employee might represent the last line of defense capable of recognizing an anomaly, questioning an unusual request, or reporting suspicious behavior that could otherwise lead to a significant security breach.[3] The inherent human capacity for intuition and the ability to discern requests or situations that seem out of context are particularly valuable in thwarting attacks that heavily rely on deception and manipulation, characteristics often missed by purely technical detection systems.[2] This dual role, as both the first sensor and the ultimate arbiter of suspicious activity, underscores the critical importance of cultivating a strong human firewall as an integral component of a comprehensive and layered cybersecurity strategy.

4. Criticisms and Limitations: Expert Skepticism Towards the "Human Firewall"

While the potential benefits of a "human firewall" are widely acknowledged, some cybersecurity experts express skepticism regarding the concept, primarily due to the inherent limitations associated with relying on human behavior for security. These experts point out that despite training and awareness initiatives, the possibility of human error remains a significant factor, making it unrealistic to consider individuals as an impenetrable "firewall."[16] The human element is consistently identified as a primary weakness in the overall security chain, with a substantial percentage of successful data breaches directly attributed to mistakes made by employees.[2] Critics emphasize the increasing sophistication of cybercriminals, who are adept at developing and deploying highly effective social engineering tactics that exploit fundamental human psychological vulnerabilities, making even well-trained individuals susceptible to manipulation.[2] A key concern raised by these experts is that an over-reliance on the "human firewall" concept might inadvertently lead organizations to neglect or underinvest in the development and maintenance of robust and essential technological security measures, creating a potentially dangerous imbalance in their overall security strategy.[20]

The effectiveness of relying on humans as a primary security mechanism is challenged by several inherent limitations and concerns. Human error remains a significant vulnerability, with employees potentially falling victim to increasingly sophisticated phishing scams, adopting weak and easily compromised passwords, or making unintentional errors in judgment that can lead to security breaches.[2] Cyber attackers frequently employ various social engineering tactics, including phishing emails designed to trick users into revealing sensitive information, pretexting scenarios where attackers impersonate trusted entities, baiting techniques that lure victims with enticing offers, and scareware tactics that induce fear to prompt harmful actions[2]. The threat from within also poses a significant risk, as malicious or simply negligent insiders can intentionally or unintentionally compromise an organization's security.[2]

Over time, users can experience security fatigue or complacency, becoming desensitized to frequent security warnings and protocols, potentially leading to lapses in vigilance.[17] Maintaining a state of constant vigilance and security awareness can be particularly challenging for individuals whose primary responsibilities lie outside of cybersecurity.[2] The ever-evolving nature of the cyber threat landscape necessitates continuous and adaptive training programs, which can be resource-intensive and may not always keep pace with the latest attack vectors.[3] Furthermore, attackers often exploit psychological principles, leveraging emotions such as urgency, fear, and curiosity to bypass rational decision-making processes and manipulate individuals into taking actions that compromise security.[19]

The significant role of human error in the vast majority of data breaches underscores the inherent vulnerabilities associated with placing excessive reliance on individuals as a primary security control. Numerous studies and reports consistently indicate that a substantial percentage of security incidents can be traced back to human actions or inactions.[5] Employees, often burdened with numerous responsibilities and facing time constraints, can be easily deceived by well-crafted malicious emails or sophisticated social engineering tactics that appear legitimate.[5] A lack of sufficient awareness or the development of a false sense of security can lead employees to believe that their organization's security measures are infallible, potentially fostering risky online behaviors.[17] It is important to recognize that even the most well-intentioned and diligently trained employees are still human and therefore susceptible to making mistakes or falling prey to highly sophisticated attacks that exploit their emotions, trust, or lack of complete contextual information.[19] This inherent potential for human fallibility highlights the critical need for organizations to adopt a balanced cybersecurity strategy that incorporates robust technical safeguards as the foundational layer of defense, rather than solely depending on the vigilance and actions of their employees.

5. Comparing and Contrasting Expert Perspectives on the "Human Firewall"

The discourse surrounding the "human firewall" in cybersecurity reveals a comparative landscape where proponents and critics present distinct yet often overlapping arguments. Those who advocate for the concept emphasize the critical need for a human layer of defense, particularly against the growing prevalence of social engineering attacks that specifically target human vulnerabilities. They rightly point out that employees are frequently the initial point of contact for various types of cyber threats and highlight the significant potential to empower these individuals through targeted training and awareness programs, transforming them into active and effective defenders of organizational assets.

Conversely, experts who express skepticism towards the "human firewall" concept primarily focus on the inherent unreliability of human behavior in the context of security. Their arguments are often grounded in the consistently high rates of human error observed in security breaches, suggesting that while security awareness is undoubtedly important, placing primary reliance on individuals as a "firewall" is inherently flawed. They contend that the primary focus of an organization's security efforts should remain on the implementation and maintenance of robust technical controls that do not depend on flawless human behavior to be effective.

It is important to note that both sides of this discussion generally concur on the fundamental importance of security awareness and training for employees within any organization. The core point of divergence lies in the degree to which humans can be considered a truly reliable and consistent "firewall" against cyber threats and, consequently, the appropriate allocation of resources and strategic focus in building a comprehensive cybersecurity posture. The debate centers on whether the term "human firewall" accurately reflects the capabilities and, more importantly, the limitations of human vigilance and behavior in the face of increasingly sophisticated cyber threats.

The central points of agreement among experts revolve around the necessity of security awareness and training for all employees. There is a broad consensus that educating individuals about common cyber threats, particularly social engineering tactics, is a crucial step in mitigating risks. However, the primary areas of disagreement include the appropriateness and validity of the term "human firewall" itself, the extent to which human vigilance can be considered a reliable primary security layer compared to automated technological controls, and the optimal balance that organizations should strive for between human-centric and technology-centric cybersecurity strategies. Ultimately, the contrasting perspectives underscore the critical need for organizations to adopt a balanced and layered approach to security. This approach should strategically leverage the inherent strengths of both a well-informed and vigilant workforce and the robust capabilities of technological security controls, while simultaneously acknowledging the inherent limitations and potential vulnerabilities associated with each.

6. The "Human Firewall" in Action: Examples of Successes and Failures

Numerous instances demonstrate the positive impact of successfully implemented "human firewall" strategies on an organization's security. For example, employees who are trained to recognize and promptly report suspicious emails play a crucial role in strengthening an organization's overall threat detection capabilities.[3] Encouraging and enabling staff to create strong, unique passwords for their accounts and to utilize multi-factor authentication significantly enhances the security of sensitive information and reduces the risk of unauthorized access.[3] Similarly, when employees are educated about the dangers of unauthorized software downloads and adhere to policies that restrict such activities, the risk of malware infections and subsequent system compromise is significantly reduced.[3] Organizations that invest in comprehensive and engaging training programs have reported tangible improvements in their security posture, such as a substantial increase in the rate at which employees report suspicious activity and a notable decrease in the rate at which they fall victim to simulated phishing attacks.[3] Real-world examples further illustrate the effectiveness of a human firewall. In one instance, vigilant employees at a major bank in Southeast Asia successfully identified inconsistencies within a phishing email, promptly reporting it and thereby preventing a potentially significant security breach.[24] Another notable case involved a Tesla employee who recognized and reported a lucrative bribery attempt aimed at planting malware within the company's systems, demonstrating how employee awareness and responsible action can thwart even well-funded cybercriminal activities.[24] Implementing ongoing cybersecurity training sessions that cover topics such as recognizing phishing scams, avoiding malware downloads, and identifying malicious links are common and often effective strategies for bolstering the human firewall.[25] Establishing clear and accessible procedures for employees to report any suspicious activity they encounter and fostering a culture where questioning unusual requests is encouraged further strengthens this human layer of defense.[25]

Despite the documented successes, there are also numerous instances where reliance on the "human firewall" has proven insufficient or has failed entirely, often leading to significant security incidents. Statistics consistently reveal that a substantial proportion of data breaches can be directly attributed to human error, such as employees inadvertently clicking on malicious links embedded in phishing emails or falling victim to sophisticated social engineering tactics [6]. Cybercriminals frequently exploit various social engineering techniques, including vishing (voice phishing) attacks conducted over the phone, the use of malware-laden email attachments designed to install malicious software upon opening, and tailgating, where unauthorized individuals gain physical access to secure areas by closely following authorized personnel.[26] Phishing attacks, in particular, remain a highly effective attack vector, successfully tricking unsuspecting employees into divulging sensitive personal or organizational information or unknowingly downloading malicious software onto their devices.[2] The adoption of weak or easily guessable passwords by employees continues to be a significant vulnerability that can lead to unauthorized access to critical systems and data.[18] The loss or theft of unencrypted laptops, smartphones, or other portable devices can provide malicious actors with direct access to sensitive company networks and data, often bypassing other security measures that rely on network perimeter defenses.[2] Organizations with inadequately trained employees are demonstrably more susceptible to a wide range of scams and attacks orchestrated by cybercriminals.[16] Even seemingly innocuous actions, such as clicking on a link in an email that appears to be legitimate, can have severe consequences, as highlighted in numerous data breach reports where such actions served as the initial point of compromise.[17] These examples of failures underscore the persistent challenge of human error in cybersecurity and emphasize the critical need for organizations to implement robust technical controls and to continuously reinforce security awareness and best practices among their employees.

7. Analyzing the Validity and Relevance of the "Human Firewall" in Contemporary Cybersecurity

Based on the analysis of expert opinions and historical context, the concept of the "human firewall" retains significant validity and relevance in contemporary cybersecurity practices. This is primarily due to the fact that social engineering attacks continue to be a prevalent and highly effective threat vector, directly exploiting inherent human vulnerabilities.[2] While it is clear that humans are not infallible and can be susceptible to deception and error, a well-trained and security-aware workforce provides a crucial and often indispensable layer of defense that effectively complements the array of technological security solutions deployed by organizations.[2] The emphasis on the human element signifies a necessary and positive shift towards a more holistic cybersecurity strategy, one that acknowledges that technology alone cannot provide complete protection against all forms of attack, particularly those that rely on manipulating human behavior.[2]

However, it is worth considering whether the term "firewall" accurately reflects the nature and capabilities of this human-centric security layer. The term "firewall" typically implies a consistent, rule-based, and largely automated system of protection. Human behavior, by its very nature, is subject to variability, emotions, and the potential for error, making the analogy somewhat imperfect. Alternative terms, such as "human sensor network" or "security-aware workforce," might more accurately convey the active and responsive role of individuals in identifying and reporting potential threats, rather than implying a purely preventative and impermeable barrier.

In the face of increasingly sophisticated cyber threats, including the growing utilization of artificial intelligence in attack methodologies, the role of human awareness and critical thinking becomes even more paramount. As AI is increasingly leveraged by both threat actors to create more convincing and targeted attacks and by security teams to enhance detection and response capabilities, the ability of humans to identify anomalies and contextual inconsistencies that AI might overlook remains a vital asset.[11] Human awareness is particularly crucial for recognizing and reporting novel and evolving attack methods that may not yet be cataloged or detectable by automated security systems.[3] Consequently, cybersecurity training programs must continuously adapt to address these emerging threats, including sophisticated techniques like deepfake impersonation and increasingly elaborate business email compromise (BEC) schemes designed to deceive even discerning individuals.[24] A "human-first" approach to cybersecurity underscores the importance of ensuring that individuals within an organization not only understand the potential threats but also comprehend how the security technologies deployed work and how to effectively utilize them in conjunction with their own vigilance and awareness to maintain a strong security posture.[14]

8. Conclusion and Recommendations

In conclusion, the "human firewall" concept emerged from the critical understanding that cybersecurity encompasses not only technological defenses but also the vigilance and actions of individuals within an organization, particularly as social engineering attacks became a dominant threat. While experts broadly agree on the fundamental importance of security awareness and training for employees, there are varying perspectives on the extent to which humans can be considered a consistently reliable "firewall" due to the inherent potential for human error. Despite these limitations, the concept remains highly relevant in contemporary cybersecurity. With social engineering continuing to be a significant threat vector, a well-trained and security-aware workforce provides a vital layer of defense that complements and enhances the effectiveness of technological security controls.

To effectively leverage the human element in their cybersecurity strategy while acknowledging its inherent limitations, organizations should consider the following recommendations:

- Implement continuous and engaging security awareness training programs: These programs should utilize a variety of methods, including interactive sessions, realistic simulations, and relevant real-world examples, to ensure that employees remain informed about the latest threats and best practices.[3]
- Foster a strong security-conscious culture: Create an environment where employees feel empowered and encouraged to report any suspicious activities without fear of reprisal, reinforcing the idea that security is a shared responsibility that extends across all levels of the organization.[2]
- Conduct regular phishing simulations and other security drills: These exercises are crucial for testing the effectiveness of training programs, assessing employee vigilance, and reinforcing learned security behaviors in a controlled environment.[18]
- Establish clear and easily accessible policies and procedures: Ensure that employees have access to comprehensive guidelines on cybersecurity best practices, covering areas such as password management, the appropriate handling of sensitive information, and the established protocols for reporting security incidents.[3]
- Provide employees with the necessary tools and technologies: Equip the workforce with user-friendly security tools such as password managers, multi-factor authentication mechanisms, and secure communication channels to support their role in maintaining a secure environment.[3]

- Implement robust technical security controls as the primary layer of defense: Recognize that while the "human firewall" is a valuable complementary layer, it should not be considered a replacement for essential technological safeguards such as advanced firewalls, intrusion detection and prevention systems, and endpoint security solutions.[2]
- Focus on creating a balanced cybersecurity strategy: Strive for an optimal equilibrium between human-centric security awareness initiatives and the deployment of robust technological defenses, acknowledging the inherent strengths and limitations of each approach.[11]
- Measure the effectiveness of human firewall initiatives: Utilize relevant metrics, such as phishing reporting rates, overall threat reporting rates, and employee feedback gathered through surveys, to assess the impact and identify areas for improvement in security awareness programs.[23]
- Encourage strong leadership buy-in and support: Secure visible commitment and active participation from organizational leaders to underscore the importance of security awareness and foster a culture where security is prioritized at all levels.[3]
- Adapt training and awareness efforts to address emerging threats: Continuously update training content and methodologies to effectively address new and evolving cyber threats, including sophisticated attacks that leverage artificial intelligence and other advanced techniques.[3]

By implementing these recommendations, organizations can more effectively harness the potential of their employees as a crucial layer of defense against the ever-evolving landscape of cyber threats, while also recognizing and mitigating the inherent limitations of relying solely on human vigilance.

Works cited

Can be found at https://www.c-ooda.com/book-of-five-keys

Appendix B The Evolution and Impact of "Not If, But When" in Cybersecurity Discourse

I. Introduction: Setting the Stage - The Ubiquitous "Not If, But When" in Cybersecurity

The phrase "Cybersecurity incidents are not a matter of 'if' but 'when'" has become a pervasive adage within the cybersecurity industry. It frequently appears in marketing materials, presentations by experts, and general discussions concerning the ever-present threats in the digital landscape. This statement, while intended to underscore the inevitability of cyberattacks, has also drawn criticism for its potentially fatalistic tone, prompting a search for more constructive approaches to cybersecurity communication. This report aims to investigate the history of this widely used phrase, trace its evolution within the cybersecurity domain, analyze its adoption and adaptation by various stakeholders, explore the arguments against its potentially discouraging nature, and identify alternative messaging strategies that promote a more proactive and empowering stance in the face of cyber threats. By examining the origins, usage patterns, and critiques of this phrase, this report seeks to provide a comprehensive understanding of its impact on cybersecurity discourse and explore avenues for more effective communication.

II. Genesis of the Phrase: Investigating Early Instances and Similar Phrasing

While the specific phrasing "Cybersecurity incidents are not a matter of 'if' but 'when'" gained prominence in the cybersecurity field, the underlying concept of inevitable negative events has a longer history. An early recorded use of a similar phrase, "It's not if, but when," dates back to an 1867 English periodical referencing an Italian politician.[1] This suggests that the notion of certain events being unavoidable predates the emergence of cybersecurity as a distinct field. The cybersecurity community appears to have adopted an existing idiom to convey the increasing certainty of cyber intrusions. This pre-existing understanding of inevitability might explain the rapid acceptance and widespread use of the phrase within the industry. The core idea it conveys – that negative events are bound to occur – is not unique to the realm of digital security.

Within the context of early cybersecurity discussions, a similar sentiment, though not the exact phrasing, was conveyed by prominent figures. In 2012, then FBI Director Robert Mueller reportedly stated, "There are only two types of companies; those who have been hacked and those who will be."[2] This statement, while differing in its specific wording, delivers a very similar message of inevitability concerning cyberattacks. The fact that a high-ranking law enforcement official like the FBI Director articulated this view suggests that the concept of cyberattacks being unavoidable was gaining traction and being communicated by influential voices even before the precise "not if, but when" phrasing became dominant. Mueller's quote might have laid the groundwork for the later adoption of the more concise and impactful phrasing. This narrative around inevitability further evolved with an updated version of Mueller's quote: "There are only two types of companies: those that have been hacked and those that don't know they have been hacked."[3] This evolution emphasizes the potential for breaches to occur and remain undetected, further reinforcing the idea that experiencing a cyber incident is not a matter of chance but a matter of time, and potentially, current awareness. This subtle shift in the narrative adds a layer of urgency, implying that organizations should not be complacent even if they have not yet detected an attack, suggesting a more sophisticated understanding of the threat landscape where intrusions can be silent and persistent.

The specific phrasing "it's not if you get breached, but when" is also attributed to a former FBI director, likely Robert Mueller, in a YouTube video transcript discussing cybersecurity.[4] This reinforces the idea that a prominent figure in law enforcement played a role in popularizing this type of messaging within the cybersecurity domain. Such pronouncements from individuals in positions of authority likely carried significant weight and contributed to the widespread acceptance of this perspective within the industry. Examining other early mentions in publications or speeches reveals that by 2016, the specific phrasing was in use in educational and professional contexts, as evidenced by a webinar titled "Cybersecurity: It's not a matter of "if" but "when" there will be a breach."[5] This indicates that the exact phrase was circulating within the cybersecurity community and being used to frame discussions about organizational risk well before it became a ubiquitous industry catchphrase.

III. The Role of Law Enforcement: Examining Mentions by FBI Directors Comey and Mueller

While the user's initial query suggests that former FBI Director James Comey might have been the first to use the phrase "Cybersecurity incidents are not a matter of 'if' but 'when'," a review of available materials does not directly support this attribution. Snippets from various speeches and interviews by Comey [6] highlight his focus on the increasing threat from nation-state actors, the critical need for enhanced collaboration between the private sector and the FBI in addressing cyber intrusions, and the "epidemic proportions" of cybercrime. While Comey's rhetoric consistently emphasized the pervasive and serious nature of cyber threats, aligning with the underlying message of inevitability, these sources do not contain the specific phrase in question. His strong warnings about the growing sophistication and frequency of attacks likely contributed to the overall sense of urgency within the cybersecurity community and the acceptance of the idea that breaches are highly probable, even without using the exact "not if, but when" formulation.

In contrast, the research material explicitly indicates that former FBI Director Robert Mueller used a very close variation of the phrase. One source states, "Moreover, as FBI Director Robert Mueller highlighted in 2012, it is not a matter of if you will be attacked, but when."[2] This directly confirms that Mueller employed this type of messaging as early as 2012. His broader communication strategy consistently underscored the seriousness and increasing likelihood of cyberattacks, emphasizing the growing cyber threat, the involvement of nation-states, and the necessity of robust collaboration between government agencies and the private sector.[11] Therefore, while James Comey undoubtedly contributed to raising awareness about cyber threats, the available evidence points to Robert Mueller as a key figure in popularizing the sentiment, if not the exact phrasing, of "not if, but when" within the cybersecurity landscape.

IV. Industry Embrace: How Cybersecurity Companies Have Adopted and Evolved the Statement

The sentiment that cyberattacks are inevitable has been widely adopted and adapted by cybersecurity companies and experts in their communication strategies. IT Governance USA, for example, directly uses the phrase "Cyber incidents are a matter of when, not if" in their content.[14] This demonstrates how the exact phrasing has been integrated into the messaging of organizations offering cybersecurity services and products. Experts and thought leaders in the field have also embraced this perspective. Professor Kamal Bechkoum, head of Business and Technology at the University of Gloucestershire, is quoted as saying, "In the first instance understand that a cyber-attack on your organisation is inevitable. It's really not a question of 'if', but 'when.'"[15] The use of the phrase by academics and industry commentators further solidifies its acceptance and dissemination within the cybersecurity community.

The phrase frequently appears in discussions and warnings within the business and cybersecurity sectors. At an Alliance MBS event, a panelist warned, "Companies have done a lot of things right, but it is not a matter of if but when they will come under attack."[16] This highlights its role in shaping the narrative around cybersecurity risk for businesses. Furthermore, the title of an article in Applied Radiology, "Cyberattacks: Not a Matter of If, but When," demonstrates the use of the phrase across different industries, including healthcare.[17] This indicates its perceived relevance across various sectors facing cyber threats. Palo Alto Networks, a major cybersecurity vendor, states that "The general consensus among industry experts is that an organization facing a cybersecurity breach or attack is not a matter of "if," but rather "when.""[18] This confirms the widespread acceptance of the sentiment as a "general consensus" within the cybersecurity industry.

Consulting firms like Oliver Wyman have also adopted the phrase, using it in the title of their insights, such as "Cyberattack: Not If, But When."[19] This indicates its role in framing the discussion around cyber risk for business leaders. Even individuals with direct experience in cybersecurity within law enforcement echo this sentiment. Tim Gallagher, a former FBI special agent in charge, stated, "Everybody's going to get hit."[20] While not the exact phrase, this reinforces the pervasive belief in the inevitability of attacks among cybersecurity professionals. The phrase is also commonly used in the context of data breaches, as seen in the title of a report by RWK Goodman: "Data Breach: When, Not If."[21] This shows how the general idea of inevitable cyber incidents is often narrowed down to the specific concern of data breaches for many organizations. The Oklahoma Bar Association uses the heading "NOT IF, BUT WHEN" in an article discussing cyber-attacks and the need for incident response plans.[22] This indicates the phrase's adoption even within legal and professional organizations discussing cybersecurity risks. Organizations focused on corporate governance and director education, such as the Australian Institute of Company Directors (AICD), also use the phrase in their materials, as in the title "It's not if but when a cyber security attack will happen."[23] This highlights its importance at the board level, suggesting that the inevitability of cyberattacks is a key message being communicated to business leaders. The widespread and consistent use of the "not if, but when" statement and its variations across different segments of the cybersecurity landscape underscores its entrenchment as a core tenet in the industry's understanding of cyber risk.

V. A Critical Examination: Arguments Against the Fatalistic Nature of the Phrase

Despite its widespread use, the "not if, but when" statement has faced criticism for its potentially fatalistic nature and its limitations in fostering a proactive security mindset. A YouTube video transcript discussing this very phrase highlights the concern that starting with "it's not if we get breached, but when" can be counterproductive, potentially discouraging executives from investing in necessary security initiatives.[4] The argument is that if leaders are presented with a seemingly unavoidable outcome, they might be less inclined to allocate resources towards prevention, leading to underfunding of crucial security measures. This messaging can inadvertently create a sense of helplessness, undermining efforts to improve an organization's security posture. Furthermore, the speaker in the video suggests that the fatalistic nature of the statement might lead to an overemphasis on cyber insurance as a primary mitigation strategy, rather than focusing on proactive security measures aimed at preventing breaches in the first place. If attacks are perceived as inevitable, organizations might prioritize financial recovery after an incident over investing in stronger defenses to prevent the incident from occurring.

The "not if, but when" sentiment has also been critiqued for potentially overstating the certainty of an attack for every single organization. An article in Domestic Preparedness discusses the "not if, but when" fallacy in the context of active shooter preparedness, arguing that while the phrase might be well-intentioned, it can be misleading about the likelihood of an event at a specific location.[1] Applying this logic to cybersecurity, while cyberattacks are increasingly common, the likelihood of a significant breach can vary considerably depending on an organization's size, industry, and the robustness of its security measures. The "not if, but when" statement might create a uniform sense of imminent threat that doesn't always accurately reflect the varying levels of risk across different organizations. In response to these criticisms, the speaker in the aforementioned YouTube video proposes reframing the narrative to focus on resilience rather than solely on the inevitability of breaches.[4] This alternative approach acknowledges the possibility of cyber incidents but emphasizes an organization's ability to withstand and recover from them, shifting the focus from a passive acceptance of attacks to an active preparation for them, fostering a sense of control and empowerment.

VI. Moving Beyond Fatalism: Exploring Proactive and Empowering Messaging Alternatives

Recognizing the potential drawbacks of the "not if, but when" statement, the cybersecurity field has explored alternative messaging strategies that aim to be more proactive and empowering. One prominent alternative focuses on the concept of "resiliency," emphasizing the importance of detection, response, and limiting the damage caused by cyber incidents.[4] This approach acknowledges the likelihood of breaches but centers on an organization's ability to minimize their impact, reframing the conversation from a passive acceptance of attacks to an active preparation for them, thereby fostering a sense of control and empowerment. This shift in focus is also highlighted by the Australian Institute of Company Directors (AICD), which emphasizes the need to move from "pure prevention to detection and response planning" with the ultimate goal of becoming "resilient organisations that can bounce back quickly from attacks."[23] This suggests that a balanced approach that includes prevention, detection, and response is more effective than solely concentrating on prevention under the assumption of inevitability.

Beyond the concept of resilience, messaging that highlights actionable steps empowers individuals and organizations to take control of their security posture. Recommendations such as using encrypted messaging applications, enabling multi-factor authentication, and practicing good password hygiene[24] provide concrete ways for users to mitigate risks. By offering specific, implementable advice, this type of messaging shifts from a sense of impending doom to a sense of agency and the ability to influence security outcomes. The concept of "proactive cybersecurity" has also gained traction, emphasizing prevention, continuous monitoring, threat detection, and comprehensive employee training programs.[29] Focusing on these proactive measures communicates a sense of control and the ability to actively defend against threats, rather than passively waiting for an attack to occur. This approach encourages taking initiative and implementing strategies to reduce both the likelihood and the potential impact of cyber incidents, fostering a more optimistic and action-oriented mindset. Strategies for achieving "cybersecurity success" further emphasize this proactive stance, including regular training sessions, the development of clear and concise security policies, encouraging leadership engagement in security practices, establishing incident reporting mechanisms, and actively seeking feedback to continuously improve security measures.[34] This holistic approach focuses on building a strong security culture through active participation and continuous improvement, emphasizing a shared responsibility for maintaining a secure environment.

VII. Contextual Usage Today: Analyzing the Target Audience and Intended Message

The "not if, but when" phrase is typically employed in specific contexts depending on the target audience and the intended message. When addressing business leaders and executives, as seen in materials from ChiefExecutive.net, AICD, and Oliver Wyman, the intended message is often to underscore the critical importance of taking cybersecurity seriously at the highest levels of the organization and allocating the necessary resources to address the ever-present threat.[19] By highlighting the inevitability of cyberattacks, the messaging aims to overcome potential complacency among leadership and drive investment in both preventative security measures and comprehensive preparedness plans. Cybersecurity vendors frequently utilize the phrase in their marketing materials, as exemplified by IT Governance USA and Palo Alto Networks.[14] In this context, the intended message is likely to create a sense of urgency and emphasize the necessity of the vendor's products or services to help organizations effectively prepare for the inevitable "when" an attack occurs. By stressing the inevitability, these vendors position themselves as essential partners in mitigating the potential impact of an eventual breach.

Educational content and expert discussions also commonly employ the "not if, but when" phrase, as seen in materials from Alliance MBS and Applied Radiology.[16] In these settings, the goal is often to establish a realistic understanding of the current threat landscape and the ongoing need for constant vigilance and preparedness. By acknowledging the high likelihood of cyberattacks, educators and experts aim to move the conversation beyond basic prevention strategies and towards a more comprehensive and adaptive security posture. The specific nuance of the message is often tailored to the intended audience. When communicating with technical audiences, the discussion might quickly transition to specific defensive strategies and incident response protocols after acknowledging the inevitability of an attack. Conversely, when addressing non-technical audiences, the emphasis might be on the broader business and organizational implications of cyber incidents and the importance of fostering a security-aware culture throughout the entire organization. Therefore, while the "not if, but when" statement serves as a common starting point, the specific message and subsequent discussion are often adapted based on the audience's level of understanding, their specific concerns, and the desired call to action.

VIII. Variations and Related Phrases in Cybersecurity Discourse

Over time, the original "Cybersecurity incidents are not a matter of 'if' but 'when'" statement has spawned several variations and related phrases within cybersecurity discourse. These include:

- "It's not if you get breached, but when."[4]
- "Cyber incidents are a matter of when, not if."[14]
- "Data breach: When, not if."[21]
- "It's not a matter of if but when they will come under attack."[16]
- "The question organizations are facing is not if a cyberattack will happen, but when."[19]
- Robert Mueller's quote: "There are only two types of companies; those who have been hacked and those who will be."[2]
- The evolved version of Mueller's quote: "There are only two types of companies: those that have been hacked and those that don't know they have been hacked."[3]
- "Everybody's going to get hit."[20]
- "A data breach is a question of when, not if."[21]

The subtle differences in these variations can have specific implications for the message being conveyed. For instance, phrases like "data breach: when, not if" or "it's not if you get breached, but when" focus specifically on the compromise of data, tailoring the message to the particular concern of data security and privacy. This specialization allows for more targeted communication about specific risks. The evolution of Robert Mueller's quote to include the idea that some companies "don't know they have been hacked" highlights the increasing sophistication of cyberattacks and the potential for them to go undetected for extended periods. This reflects a deeper understanding of the threat landscape and the significant challenges associated with threat detection, emphasizing the need for continuous monitoring and proactive threat hunting capabilities.

IX. Conclusion: Synthesizing Findings and Offering a Balanced Perspective on Cybersecurity Communication

The phrase "Cybersecurity incidents are not a matter of 'if' but 'when'" has become a cornerstone of cybersecurity communication, with its origins likely tracing back to similar sentiments expressed in other fields and popularized within cybersecurity by figures like former FBI Director Robert Mueller as early as 2012. The phrase and its variations have been widely adopted and adapted by cybersecurity companies, experts, and various organizations to underscore the increasing inevitability of cyberattacks across all sectors.

While the phrase effectively conveys the serious and pervasive nature of cyber threats, concerns have been raised about its potential to foster a sense of fatalism, potentially discouraging investment in preventative measures and leading to an overreliance on reactive strategies like cyber insurance. The analysis suggests a growing recognition of these limitations, with a discernible shift towards more proactive and empowering messaging strategies that emphasize resilience, continuous monitoring, threat detection, and the importance of building a strong security culture through education and engagement.

Moving forward, a balanced approach to cybersecurity communication appears most effective. While acknowledging the high likelihood of cyber incidents is crucial for driving awareness and prioritizing security, this message should be coupled with a strong emphasis on proactive measures, the development of robust incident response plans, and the cultivation of organizational resilience. By empowering organizations and individuals with actionable strategies and fostering a sense of control over their security posture, the cybersecurity community can move beyond a potentially discouraging narrative of inevitability towards a more constructive and impactful approach to mitigating cyber risks.

Table 1: Timeline of Key Mentions of "Not If, But When" and Similar Phrases

Year	Source (Speaker /Publication)	Exact Phrase or Similar Sentiment	Context/Significance
1867	English Periodical	"It's not if, but when"	Earliest recorded use of the phrase in a general context [1]
2012	Robert Mueller, FBI Director	"There are only two types of companies; those who have been hacked and those who will be"	Early articulation of the inevitability of cyberattacks by a prominent figure [2]
2016	Webinar Title	"Cybersecurity: It's not a matter of "if" but "when"	Early use of the specific phrasing in an educational context [5]

		there will be a breach"	
2020	Professor Kamal Bechkoum, University of Gloucestershire	"It's really not a question of 'if', but 'when'"	Expert opinion emphasizing the inevitability of cyberattacks [15]
2021	Palo Alto Networks	"Not a matter of "if," but rather "when.""	Statement reflecting industry consensus on the inevitability of breaches [18]

Year	Source (Speaker/Publication)	Exact Phrase or Similar Sentiment	Context/Significance
1867	English Periodical	"It's not if, but when"	Earliest recorded use of the phrase in a general context [1]
2012	Robert Mueller, FBI Director	"There are only two types of companies; those who have been hacked and those who will be"	Early articulation of the inevitability of cyberattacks by a prominent figure [2]

2016	Webina r Title	"Cybersecurit y: It's not a matter of "if" but "when" there will be a breach"	Early use of the specific phrasing in an educational context [5]
2020	Profess or Kamal Bechko um, Univers ity of Glouces tershire	"It's really not a question of 'if', but 'when'"	Expert opinion emphasizing the inevitability of cyberattacks [15]
2021	Palo Alto Networ ks	"Not a matter of "if," but rather "when.""	Statement reflecting industry consensus on the inevitability of breaches [18]

Table 2: Comparison of "Not If, But When" Messaging vs. Proactive
Alternatives

Messaging Strategy	Core Message	Potential Impact	Examples (from research snippets)
"Not If, But When"	Cyberattack s are inevitable. Focus on preparing	Can drive awareness but may also lead to fatalism	"Cyber incidents are a matter of when, not

	for the aftermath.	and discourage preventative investment.	if" [14], "Everybody's going to get hit" [20]
Resilience-Focused	Attacks may occur, but we can minimize their impact through detection, response, and recovery.	Empowers organizations to prepare actively and build capabilities to withstand attacks.	Shift from pure prevention to detection and response planning [23], Focus on being resilient [4]
Proactive Cybersecurity	We can take concrete steps to prevent attacks and reduce our vulnerability.	Fosters a sense of control and encourages the implementation of preventative measures.	Using multi-factor authentication [29], Continuous threat detection [32], Employee training [30]
Security Culture Building	Cybersecurity is a shared responsibility. Engagement and continuous improvement are key to success.	Creates a holistic approach to security involving people, processes, and technology.	Regular training sessions [34], Clear and concise policies [34], Incident reporting mechanisms [34]

Works cited

Can be found at https://www.c-ooda.com/book-of-five-keys

Appendix C The Impact of Generative and Agentic AI on Security Automation and Orchestration in Cybersecurity

1. Executive Summary

Security Automation and Orchestration (SAO) has become a cornerstone of modern cybersecurity, providing organizations with the means to manage the increasing volume and sophistication of cyber threats. By integrating disparate security tools and automating repetitive tasks, SAO aims to enhance efficiency, accelerate incident response, and minimize human error. The emergence of generative and agentic artificial intelligence (AI) presents a transformative opportunity to further evolve SAO capabilities. Generative AI, with its ability to create new content and insights from data, can automate threat intelligence analysis, generate security content, and summarize incidents. Agentic AI, characterized by its autonomy and decision-making capabilities, can enable autonomous threat detection, intelligent response actions, and adaptive security controls. While the integration of these AI technologies into SAO promises significant benefits, it also introduces challenges related to complexity and the need for specialized expertise. This report explores the core principles and objectives of SAO, the advantages and disadvantages of its adoption, the definitions and functionalities of generative and agentic AI in cybersecurity, their potential applications within SAO, and the perspectives of Torq.io on this evolving landscape. Ultimately, the synergy between AI and SAO is poised to reshape the future of cybersecurity, leading towards more proactive, efficient, and resilient security operations.

2. Defining Security Automation and Orchestration (SAO) in Cybersecurity

2.1. Core Principles of SAO:

The foundation of Security Automation and Orchestration lies in three core principles: integration, coordination (orchestration), and automation. Integration is the act of connecting various security tools and technologies, both those specifically designed for security and others that are not, to enable them to function as a unified system.[1] This connectivity often involves utilizing Application Programming Interfaces (APIs) and custom connectors to facilitate the exchange of data and the coordination of actions between diverse systems, such as firewalls, network monitoring tools, antivirus software, and endpoint security solutions.[3] The capacity to integrate a broad spectrum of security technologies is essential for SAO, allowing organizations to maximize the value of their existing security investments. However, this also presents a potential challenge when dealing with older systems or tools that lack robust integration capabilities. Without effective integration, the creation of automated workflows that span multiple systems becomes impossible, as these connections form the fundamental infrastructure for any orchestration efforts.

Coordination, or orchestration, involves the strategic arrangement and sequencing of different security tasks across a variety of tools and technologies to establish cohesive and goal-oriented security workflows.[3] Security orchestration defines the logical progression of a security plan, encompassing the stages of incident identification, thorough analysis, effective response, and complete recovery, ensuring that all involved tools operate in a synchronized manner.[2] This principle focuses on creating workflows that initiate automations to interact with each other, carefully determining the specific timing and manner of these interactions.[5] Orchestration differs from simple automation by emphasizing the overall process flow within security operations. It necessitates a deep understanding of established incident response procedures and the specific capabilities of the various security tools being utilized. While automating individual tasks can provide localized benefits, orchestration provides the essential context and flow, ensuring that these automated tasks collectively contribute to achieving broader security objectives.

Automation, the third core principle, centers on the use of automated tools and predefined processes to execute specific security tasks, often those that are repetitive, without the need for direct human intervention.[1] Examples of such tasks include the automated deployment of critical security patches, the initial investigation of routine security incidents, and the consistent implementation of defined security controls.[1] Automation streamlines routine activities like the systematic collection of data, the confirmation of security incidents based on predefined criteria, and the execution of initial response measures, thereby freeing up valuable time and resources for security teams.[4] This principle aims to increase operational efficiency and significantly reduce the workload on security analysts, enabling them to concentrate their expertise on more complex and strategically important tasks. However, it is crucial to ensure that only well-defined and thoroughly tested processes are automated to avoid any unintended negative consequences. Automating repetitive manual tasks offers significant time savings and reduces the likelihood of human error, allowing human analysts to apply their specialized knowledge to situations where it is most needed.

2.2. Key Objectives of SAO:

The implementation of SAO in cybersecurity operations is driven by several key objectives, primarily aimed at enhancing an organization's ability to effectively manage and respond to cyber threats. One of the primary objectives is enhancing efficiency within security operations.[1] This is achieved by streamlining security operations teams' workflows through the automation of repetitive tasks, which significantly reduces manual effort and enables a much faster response to emerging security incidents.[1] Furthermore, SAO aims to centralize security data and various security functions within a unified platform, which greatly expedites the processes of investigation and decision-making during security incidents.[2] By automating routine procedures such as vulnerability scanning and the generation of security reports, organizations can also substantially reduce the time and resources spent on these manual processes.[6] The pursuit of efficiency gains is a fundamental motivation for adopting SAO, as it allows organizations to effectively manage an increasing volume of sophisticated threats while operating with often limited resources. Therefore, the ability to quantify these efficiency improvements through metrics like time saved per incident becomes a critical aspect of evaluating the success of SAO implementations.

Another crucial objective of SAO is to enable faster incident response.[1] This is achieved through the automation of predefined workflows, often referred to as playbooks, which guide the response to common security incidents like phishing attacks or malware infections.[8] By automating these initial responses, organizations can significantly reduce the mean time to detect (MTTD) and the mean time to respond (MTTR) to security incidents.[6] Faster incident response is paramount as it directly minimizes the potential impact of cyberattacks, thereby reducing the likelihood of significant damage and prolonged downtime. This objective underscores the critical importance of developing well-defined and thoroughly tested incident response playbooks as a core component of any SAO strategy. The quicker an organization can respond to and contain a security breach, the less opportunity attackers have to cause significant harm to systems and data.

Improving accuracy in security operations is also a key objective of SAO.[1] Orchestration ensures the consistent execution of predefined response actions, which significantly minimizes the risk of human error during the critical process of incident handling.[1] Furthermore, SAO contributes to improved accuracy by reducing the occurrence of false positives and the associated alert fatigue experienced by security analysts. This is achieved through the intelligent correlation of data from multiple diverse sources and the application of sophisticated automation techniques.[7] Human error remains a significant contributing factor to security breaches, and SAO directly addresses this vulnerability by standardizing security processes and automating critical actions, leading to more reliable and accurate security operations. By automating tasks, organizations reduce their reliance on manual processes, which are inherently more susceptible to mistakes, particularly when security teams are operating under pressure during an incident.

Enhancing scalability is another vital objective of SAO in the face of evolving cyber threats and expanding IT environments.[1] SAO enables organizations to effectively handle a growing volume of security alerts and incidents without requiring a proportional increase in security resources.[1] As organizations experience growth and the complexity of the threat landscape continues to increase, SAO provides the necessary scalability to maintain an effective security posture without incurring unsustainable increases in staffing levels. Manual security operations often struggle to keep pace with the sheer volume and increasing sophistication of modern cyber threats, making the scalability offered by SAO a crucial advantage.

Finally, SAO aims to provide centralized management and visibility over security operations.[1] Orchestration platforms typically offer a centralized dashboard that allows for the comprehensive management of security alerts, ongoing incidents, and various response activities, providing security teams with enhanced visibility and greater control over their security operations.[1] This often involves integrating security tools, IT operations systems, and threat intelligence platforms into a single, unified console.[10] By aggregating and correlating data originating from multiple security tools, SAO delivers a more holistic and comprehensive view of the organization's overall security environment.[13] Centralized management simplifies the complexities of security operations, improves overall situational awareness, and facilitates more effective collaboration among different security teams. This unified view also aids in identifying underlying patterns and trends across seemingly disparate security events, leading to more informed and proactive security strategies.

3. Benefits of Implementing SAO in Cybersecurity Operations

3.1. Improved Efficiency and Reduced Workload for Security Teams:

Implementing SAO yields significant improvements in the efficiency of security operations and a substantial reduction in the workload faced by security teams.[1] By automating routine and repetitive tasks, such as the systematic collection of logs, the initial triage of security alerts, the enrichment of data with contextual information, and the preliminary analysis of security incidents, SAO frees up security analysts to dedicate their time and expertise to more complex and strategically important activities. These higher-value tasks include proactive threat hunting to uncover hidden or emerging threats and conducting in-depth investigations into sophisticated cyberattacks.[1] Furthermore, SAO plays a crucial role in reducing the pervasive issue of alert fatigue, which often overwhelms security analysts, by intelligently filtering out false positive alerts and effectively prioritizing those alerts that represent genuine threats requiring immediate attention.[6] The implementation of SAO also streamlines existing workflows within security operations and enhances the level of collaboration among different security teams, leading to a more cohesive and effective security posture.[1] By automating the more mundane and repetitive aspects of their work, SAO not only increases the overall efficiency of security operations but can also lead to increased job satisfaction and potentially better retention rates for security analysts, particularly in a field that is currently facing a significant shortage of skilled cybersecurity professionals.

3.2. Faster Incident Response and Remediation:

A critical benefit of SAO implementation is the significant acceleration of incident response and remediation processes.[1] SAO enables the automation of incident response playbooks, which are predefined sets of actions designed to address specific types of security incidents. This automation allows for quick and consistent reactions to common threats such as phishing attempts, malware infections, and data breaches.[1] Consequently, the time required to detect, contain, eradicate, and ultimately recover from security threats is substantially reduced, which in turn minimizes the potential for significant damage and prolonged periods of operational downtime.[1] Furthermore, SAO facilitates the orchestration of actions across a multitude of diverse security tools, ensuring a coordinated and highly effective response to security incidents.[1] The speed at which an organization can respond to a cyberattack is a critical factor in determining the overall impact of the incident. SAO significantly accelerates this crucial process, leading to lower costs associated with potential data breaches and disruptions to business operations. A rapid and automated response can effectively prevent a seemingly minor security incident from escalating into a major organizational crisis.

3.3. Reduced Human Error and Enhanced Accuracy:

The adoption of SAO in cybersecurity operations leads to a notable reduction in human error and a significant enhancement in the overall accuracy of security processes.[1] By ensuring the consistent execution of established security processes and clearly defined policies, SAO effectively eliminates inconsistencies that can often arise due to human mistakes, inherent biases, or a lack of adequate training among security personnel.[1] Moreover, SAO automates critical tasks such as data collection from various sources, the detailed analysis of security events, and the generation of comprehensive reports, thereby significantly reducing the likelihood of manual errors occurring during these important processes.[2] Furthermore, SAO contributes to improving the accuracy of threat detection by consistently applying automated analysis techniques, ensuring that potential threats are identified more reliably and with greater precision.[1] Human error remains a significant vulnerability that cybercriminals often exploit. SAO helps to mitigate this inherent risk by automating critical security functions and ensuring the consistent and accurate application of security policies across the organization. By automating tasks, organizations reduce their dependence on manual processes, which are inherently more prone to mistakes, especially when security teams are faced with high-pressure situations during a security incident.

3.4. Improved Scalability and Centralized Management of Security Operations:

Appendix C The Impact of Generative and Agentic AI on Security
Automation and Orchestration in Cybersecurity
227

SAO provides organizations with improved scalability and enables centralized management of their security operations.[1] By automating many routine security tasks and orchestrating workflows across different security tools, SAO empowers security teams to effectively handle a significantly larger volume of security alerts and incidents without the need for a proportional increase in staffing levels.[1] This is particularly important in addressing the ongoing cybersecurity skills shortage that many organizations face. Furthermore, SAO solutions typically offer a single, centralized platform for managing and monitoring the entire security infrastructure, which significantly improves overall visibility and control over security operations.[1] This centralized approach also facilitates better collaboration and the seamless sharing of critical information among different security teams and relevant stakeholders within the organization.[8] The ability to scale security operations efficiently is crucial for organizations that are grappling with an ever-increasing number and complexity of cyber threats. SAO provides the necessary tools and established frameworks to effectively manage this growth without requiring unsustainable increases in human resources. Centralized management simplifies the inherent complexities of security operations and provides a comprehensive and holistic view of the organization's overall security posture, enabling more informed and effective decision-making.

4. Challenges and Potential Drawbacks of Adopting SAO

4.1. Integration Complexities with Existing Security Infrastructure:

One of the primary challenges associated with adopting SAO is the complexity of integrating it with an organization's existing security infrastructure.[1] Organizations often utilize a diverse array of security tools and technologies, each potentially employing different data formats, APIs, or communication protocols, making the process of ensuring seamless integration intricate and often time-consuming.[1] A significant hurdle lies in ensuring seamless collaboration and the smooth flow of critical data between all the integrated components, which is essential for the effective functioning of automated workflows and orchestrated responses.[4] The challenge becomes even more pronounced when attempting to integrate modern SAO solutions with legacy systems that rely on outdated technologies and proprietary software, which may lack compatibility with newer integration methods.[34] The overall effectiveness of an SAO implementation is heavily dependent on successful integration. Organizations need to develop a well-defined and comprehensive integration strategy and may need to acquire specialized technical expertise to effectively navigate these inherent complexities. Furthermore, the selection of an SAO platform should carefully consider its level of compatibility with the organization's current security technology stack to minimize potential integration challenges. If the various security tools within an environment cannot effectively communicate and readily share critical data, the intended benefits of orchestration and automation will be significantly limited, hindering the overall effectiveness of the SAO deployment.

4.2. The Need for Skilled Personnel for Implementation and Management:

The efficient utilization of SAO platforms necessitates the involvement of skilled personnel who possess a thorough understanding of both cybersecurity practices and the specific technical intricacies of the chosen platform.[4] Developing and effectively tailoring the necessary workflows and automated playbooks requires a significant level of expertise in both the operational aspects of security and the principles of automation technologies.[4] Moreover, certain SAO solutions may require hands-on knowledge of scripting languages, such as Python, Ruby, or Perl, for the purpose of creating custom integrations with other security tools and for building sophisticated automation playbooks to address specific organizational needs.[32] A significant challenge for many organizations is either training their existing security staff to acquire these specialized skills or recruiting qualified individuals who already possess the necessary expertise, especially given the well-documented and ongoing shortage of skilled professionals within the cybersecurity domain.[4] The lack of sufficient in-house expertise can become a major impediment to successful SAO adoption. Organizations may need to make substantial investments in comprehensive training programs for their current staff or allocate resources to hire new personnel who possess the required technical skills to effectively implement, manage, and maintain their SAO environment to realize its full potential. While SAO platforms are powerful tools designed to enhance security operations, they require skilled and knowledgeable operators to properly configure, manage, and continuously maintain them to ensure optimal performance and effectiveness.

4.3. Potential for Misaligned Expectations and Over-Automation:

Organizations embarking on adopting SAO may sometimes harbor unrealistic expectations regarding the capabilities and potential outcomes of these platforms, such as the belief that SOAR can automatically handle and resolve every conceivable security challenge or automate every single tedious task within their security operations.[22]

A significant potential pitfall lies in attempting to automate processes that are inherently flawed or have not been clearly and effectively defined, which can lead to unintended negative consequences and may not actually result in any tangible improvements in overall efficiency or operational performance.[22] Furthermore, the practice of over-automation, particularly without the establishment of appropriate levels of human oversight and intervention, can lead to critical security incidents being mishandled or even completely overlooked by the automated systems, potentially exacerbating the initial problem.[6] A strategic and carefully considered phased approach to the implementation of automation is therefore crucial.

Organizations need to meticulously identify the specific processes that are truly suitable for automation and ensure that they maintain a well-defined balance between the capabilities of automated systems and the critical need for human judgment and intervention, especially when dealing with complex or ambiguous security scenarios. Automation should be viewed as a tool to augment and enhance human capabilities within security operations, rather than a complete replacement for the critical thinking and nuanced decision-making that experienced security professionals can provide.

4.4. The Importance of Continuous Monitoring and Adaptation:

Successful SAO implementations are not static deployments but rather require ongoing and consistent monitoring, rigorous testing, and continuous refinement to ensure that they remain effective in the face of constantly evolving cyber threats and adapt appropriately to any changes within the organization's IT environment.[22] The automation playbooks that are initially developed and implemented within an SAO platform may become outdated and less effective as cyberattack tactics, techniques, and procedures (TTPs) continue to evolve and become more sophisticated.[22] Therefore, it is essential for organizations to regularly review operational metrics and key performance indicators (KPIs) that are relevant to their security operations to accurately assess the ongoing effectiveness of their SAO deployments and to identify any areas where further improvements or adjustments may be necessary.[6] SAO should not be viewed as a "set it and forget it" type of solution. To maximize its long-term value and ensure its continued effectiveness in protecting organizational assets, continuous monitoring and proactive adaptation are absolutely necessary. The threat landscape in cybersecurity is in a state of constant flux, with new threats and attack methods emerging regularly. Consequently, security automation and orchestration systems must also undergo a process of continuous evolution and improvement to maintain their effectiveness and provide ongoing value to the organization's security posture.

5. Generative and Agentic AI in Cybersecurity: Definitions and Core Concepts

5.1. Generative AI:

Generative AI represents a significant advancement in the field of artificial intelligence, focusing on the creation of models that can generate entirely new and original content.[39] This content can take various forms, including text written in natural language, realistic images, synthesized audio, computer code, and more, all derived from the underlying patterns and structures learned from the extensive datasets on which these models are trained.[43]

Unlike traditional AI models that primarily analyze existing data or make predictions based on it, generative AI learns the intricate relationships within massive datasets to produce novel outputs that closely mimic the characteristics of the original training data without simply replicating it verbatim.[43] This capability is achieved through the utilization of advanced machine learning techniques, particularly deep learning models such as generative adversarial networks (GANs), variational autoencoders (VAEs), large language models (LLMs), and transformer architectures, which enable these models to capture complex data distributions and generate coherent and contextually relevant content.[43]

Generative AI models are also highly adaptable, capable of producing different types of content based on specific prompts provided by users, and they can leverage machine learning algorithms to continuously recognize, predict, and generate content based on the diverse datasets they are granted access to.[44] The unique ability of generative AI to create new and realistic content has profound implications for the field of cybersecurity, offering opportunities to both significantly enhance existing security measures and potentially be exploited by malicious actors to create more sophisticated and evasive cyber threats. The quality, diversity, and speed at which generative AI can produce content are therefore key characteristics that define its potential impact on the cybersecurity landscape.

The functionalities of generative AI are diverse and rapidly expanding, encompassing the ability to generate human-like text, create realistic images, synthesize audio, and even produce video content.[42] Beyond content creation, generative AI can also be employed to summarize and synthesize vast amounts of information from diverse sources, making complex data more easily understandable and actionable.[45] In the realm of software development and security, generative AI can assist in generating and debugging computer code, potentially accelerating the development process and aiding in the identification of vulnerabilities.[44]

Furthermore, its capability to learn complex patterns makes it invaluable for creating realistic simulations of various cyber threats, which can be used for training security personnel and rigorously testing the effectiveness of existing security defenses.[41] Generative AI also excels at analyzing large datasets to identify subtle patterns and anomalies that may indicate the presence of security threats, enhancing an organization's ability to detect and respond to malicious activity.[40] These diverse functionalities underscore the versatility of generative AI as a powerful tool with a wide range of potential applications across various aspects of cybersecurity, from proactive threat detection to enhancing the skills and preparedness of security teams.

The relevance of generative AI to the field of cybersecurity is becoming increasingly significant, as it offers the potential to enhance threat detection and improve incident response capabilities by efficiently analyzing vast quantities of data in near real-time.[40] Generative AI can also automate many routine and time-consuming tasks that are typically performed by security analysts, allowing these professionals to focus their expertise on higher-level strategic initiatives and more complex security challenges.[40] This automation can be particularly beneficial for understaffed Security Operations Center (SOC) teams, as it helps to augment the capabilities of existing analysts and streamline critical workflows.[40] Moreover, generative AI can contribute to improving predictive analysis in cybersecurity, enabling organizations to anticipate potential future threats and more effectively manage vulnerabilities within their systems.[40] Another important application lies in its ability to generate synthetic data that closely resembles real-world data, which can be invaluable for training security models and algorithms without compromising the privacy of sensitive information.[41] Overall, generative AI has the potential to help cybersecurity teams overcome many of the persistent challenges they face, including the overwhelming volume of security data, the shortage of skilled cybersecurity professionals, and the ever-increasing sophistication of cyber threats. The capacity of generative AI to learn from data and subsequently generate new insights and content makes it a powerful and increasingly essential tool in the ongoing effort to defend against cybercrime and maintain a strong and resilient security posture.

5.2. Agentic AI:

Agentic AI represents an even more advanced paradigm within artificial intelligence, characterized by its ability to autonomously analyze information, make independent decisions, and take concrete actions based on predefined objectives, all with minimal direct human oversight.[60] These systems exhibit a high degree of autonomy, demonstrating goal-driven behavior and a remarkable capacity for adaptation, which allows them to operate effectively within dynamic and often unpredictable environments.[65] Agentic AI is capable of perceiving its surrounding environment, intelligently reasoning through complex tasks, and dynamically modifying its actions based on updated information and real-time feedback.[63] A key characteristic of agentic AI is its ability to continuously learn and adapt to changes in its environment, particularly in the context of cybersecurity, where it can refine its threat detection and incident response strategies based on insights gained from past security incidents and evolving threat patterns.[61] This level of autonomy and adaptability distinguishes agentic AI from traditional forms of automation, enabling it to function more like an intelligent and autonomous security analyst, capable of handling a wide range of security tasks without constant human intervention.

The functionalities of agentic AI in cybersecurity are designed to enable proactive and autonomous threat management. One of its defining functionalities is the ability to make autonomous decisions and initiate responses to detected security threats.[60] Upon identifying a potential threat, an agentic AI system can immediately take actions such as isolating compromised systems to prevent further spread, blocking malicious network access to contain the attack, or alerting human security teams to escalate the incident for more in-depth analysis.[60] To achieve this, agentic AI systems are equipped with the capability for continuous data collection and monitoring, gathering information from diverse sources across the IT environment, including network traffic patterns, endpoint activity logs, user behavior analytics, and external threat intelligence feeds.[62] Instead of relying solely on predefined rules, agentic AI utilizes sophisticated behavioral analysis techniques to identify suspicious activities and subtle deviations from established baseline patterns of normal behavior.[62] This advanced analytical capability allows agentic AI to effectively detect novel and previously unseen threats, such as zero-day exploits and advanced persistent threats (APTs), which might otherwise evade detection by traditional security systems that depend on known signatures or rule-based detection mechanisms.[62] Furthermore, agentic AI can perform autonomous remediation of identified threats and potential risks in real-time, taking immediate steps to neutralize the danger and restore systems to a secure state.[61] Agentic AI is also capable of intelligent alert investigation, automatically summarizing the key aspects of security alerts, and prioritizing them based on their potential severity and impact, thereby significantly reducing the problem of alert fatigue that often burdens security analysts.[64] These functionalities collectively demonstrate the potential of agentic AI to revolutionize cybersecurity by providing a more proactive, autonomous, and ultimately more effective approach to defending against the ever-evolving landscape of cyber threats.

In the context of cybersecurity, agentic AI exhibits several distinctive features that set it apart from traditional security tools and even other forms of AI. One of its most notable features is its ability to operate independently, without requiring constant human input or direction, effectively functioning as an autonomous security analyst capable of making real-time decisions.[61] Agentic AI systems are specifically designed with clear security objectives in mind and are capable of autonomously planning and executing the necessary steps to achieve those objectives with minimal human intervention.[61] A crucial aspect of agentic AI in cybersecurity is its capacity for continuous learning and adaptation to the ever-changing threat landscape.[61] By constantly analyzing new data and observing the outcomes of its actions, agentic AI can improve its ability to accurately detect and effectively respond to novel and emerging cyberattacks, ensuring that security defenses remain current and relevant. This autonomous and adaptive nature of agentic AI has the potential to significantly reduce incident response times, a critical factor in minimizing the damage caused by successful cyber intrusions, and to enhance the overall security posture of an organization by providing a more proactive and less reactive approach to security management.[60] Unlike traditional security tools that operate based on predefined rules and require human intervention for complex decision-making, agentic AI can think and act independently, making it a more powerful and versatile defense mechanism against the increasingly sophisticated tactics employed by modern cyber adversaries.

6. Applications of Generative AI in Security Automation and Orchestration

6.1. Automated Threat Intelligence Analysis and Enrichment:

Generative AI holds significant promise for automating the analysis and enrichment of threat intelligence within security operations.[40] These AI models can efficiently process vast quantities of threat data originating from diverse sources, including specialized threat intelligence feeds, security-focused blogs, and academic research papers, to identify emerging threats, recurring patterns in attacks, and critical indicators of compromise (IOCs) that security teams can use to proactively defend their systems.[40] Furthermore, generative AI can be utilized to generate concise and informative summaries of complex threat intelligence reports, making the often technical and detailed information more accessible and readily understandable for security analysts, regardless of their specific area of expertise.[45] This capability extends to enriching existing security alerts with valuable contextual information derived from various threat intelligence sources. By automatically adding relevant details about the nature of the threat, its potential impact, and known mitigation strategies, generative AI can significantly improve the prioritization of security cases and enhance the effectiveness of incident response efforts.[40] Moreover, by analyzing historical patterns in cyberattacks and identifying emerging trends, generative AI can contribute to predicting potential future threats and likely attack vectors, allowing organizations to proactively implement preventative measures and strengthen their defenses before an actual attack occurs.[40] The automation of threat intelligence analysis and enrichment through generative AI can dramatically improve the efficiency and overall effectiveness of security operations, enabling security teams to be more proactive in anticipating and mitigating potential cyberattacks in an increasingly complex threat landscape.

6.2. Generation of Security Content (e.g., reports, policies, training materials):

Generative AI offers a powerful capability to automate the creation of various forms of security content, leading to significant time savings and ensuring a higher degree of consistency across different types of documentation.[40] For instance, it can be used to automatically generate comprehensive security reports, such as detailed incident reports following a security breach, thorough compliance reports required for regulatory adherence, and in-depth vulnerability assessment reports that highlight potential weaknesses in systems.[40] Furthermore, generative AI can assist in the creation of organizational security policies and procedures by leveraging best practices and incorporating specific requirements unique to the organization, ensuring that these foundational documents are both comprehensive and up-to-date.[41] Another valuable application lies in the generation of realistic and engaging security awareness training materials, such as simulated phishing emails designed to test employee vigilance and educational content aimed at improving overall security awareness and reducing the likelihood of human error leading to security incidents.[41] Generative AI can even be employed to generate snippets of computer code or entire programs that are designed to automate specific security tasks or implement necessary security controls within the IT infrastructure.[44] By automating the creation of these diverse forms of security content, organizations can ensure that their employees and security teams have access to the necessary information and tools to maintain a strong and resilient security posture, while also freeing up valuable time for security professionals to focus on more strategic and complex tasks that require human expertise and critical thinking.

6.3. Incident Summarization and Reporting:

In the critical area of incident response, generative AI can play a vital role in significantly enhancing the efficiency of incident summarization and reporting.[40] When a security incident occurs, it often generates a large volume of data from various security systems, including Security Information and Event Management (SIEM) platforms, Endpoint Detection and Response (EDR) solutions, and network traffic analysis tools. Generative AI can quickly process and synthesize this diverse data to provide concise summaries of the key details of the incident.[40] By extracting the most relevant information and presenting it in an easily digestible format, generative AI enables security analysts to rapidly understand the nature of the security event, its potential scope, and the initial steps that have been taken or need to be taken to contain and remediate the threat.[40] This accelerated understanding of security incidents allows security teams to focus their efforts more effectively on the most critical aspects of the response, ultimately leading to faster resolution times and a reduction in the overall impact of the incident. Furthermore, generative AI can automate the generation of comprehensive incident reports, which are essential for both internal communication with stakeholders and for meeting various compliance and regulatory requirements.[40] By automating this often time-consuming process, generative AI frees up security analysts to concentrate on the more technical and strategic aspects of incident management, such as in-depth analysis of the attack's origin, identification of affected systems, and development of effective long-term prevention strategies.

6.4. Potential for Enhanced Predictive Capabilities and Vulnerability Management:

Generative AI demonstrates significant potential for enhancing predictive capabilities and improving vulnerability management processes within cybersecurity operations.[40] By analyzing vast amounts of historical vulnerability data in conjunction with real-time threat intelligence, generative AI models can identify patterns and correlations that may indicate potential future vulnerabilities within an organization's systems and predict likely attack vectors that malicious actors might exploit.[40] This proactive insight allows organizations to anticipate potential security weaknesses and take preventative measures before they can be leveraged by attackers. Additionally, generative AI can be employed to recommend or even automatically deploy necessary security patches and software updates based on the analysis of identified vulnerabilities, ensuring that systems are promptly secured against known weaknesses.[40] Furthermore, its ability to create realistic simulations of various types of cyberattacks can be invaluable for proactively identifying potential weaknesses and blind spots in an organization's existing security defenses.[41] By leveraging these predictive capabilities, generative AI can help organizations transition from a reactive security posture to a more proactive one, allowing them to stay ahead of potential threats and significantly strengthen their overall security resilience. The ability to anticipate future vulnerabilities and proactively address them represents a significant advantage in the ongoing battle against cybercrime.

7. Applications of Agentic AI in Security Automation and Orchestration

7.1. Autonomous Threat Detection and Anomaly Recognition:

Agentic AI stands out for its capacity to autonomously detect threats and recognize anomalies within cybersecurity environments.[61] These intelligent systems can continuously monitor network activity, analyze endpoint behavior, and scrutinize user actions in real-time to identify any deviations from established baselines or suspicious patterns that may indicate a potential security threat, all without requiring direct human intervention.[61] By leveraging sophisticated behavioral analysis techniques, agentic AI can effectively identify subtle anomalies and suspicious activities that might not trigger traditional rule-based security systems, enabling the detection of novel and advanced threats such as zero-day exploits and advanced persistent threats (APTs).[62] These systems are designed to analyze vast amounts of data at machine speed, allowing them to pinpoint subtle indicators of malicious activity that might be easily overlooked by human analysts or conventional security tools.[64] The ability of agentic AI to autonomously detect threats and recognize anomalies in real-time significantly enhances an organization's overall security posture by enabling a more proactive and less reactive approach to identifying and responding to potential cyberattacks. Unlike traditional security systems that rely on predefined rules and known signatures, agentic AI's capacity for continuous learning and adaptation allows it to identify and flag previously unseen threats, providing a critical advantage in the ever-evolving landscape of cybersecurity.

7.2. Intelligent and Automated Incident Response Actions (Containment, Remediation):

A key application of agentic AI in security automation and orchestration lies in its ability to perform intelligent and automated incident response actions.[60] Upon the autonomous detection of a security threat, agentic AI systems can automatically initiate containment measures to prevent the threat from spreading further within the organization's network.[60] This can involve actions such as isolating compromised systems from the rest of the network, blocking malicious network traffic originating from or destined to known threat actors, or terminating malicious processes that are actively running on infected machines.[60] Furthermore, agentic AI can orchestrate complex, multi-step response workflows that span across various security tools within the organization's ecosystem, all without requiring direct human intervention at each step.[62] This includes the automation of predefined incident response playbooks, which are triggered based on the specific type and assessed severity of the security incident that has been detected.[61] The ability of agentic AI to autonomously execute these critical response actions in real-time significantly reduces the time it takes to contain and remediate security incidents, thereby minimizing the potential for widespread damage and prolonged disruptions to business operations. This rapid and automated response capability is particularly crucial in dealing with sophisticated and fast-moving cyberattacks where every second counts in mitigating the overall impact.

7.3. Adaptive Security Controls and Policy Enforcement:

Agentic AI enables the implementation of adaptive security controls and the autonomous enforcement of security policies within an organization's cybersecurity framework.[61] These intelligent systems can dynamically adjust security rules and policies in real-time, taking into account evolving attack patterns that are being observed across the threat landscape, as well as the specific risk posture that the organization has defined for its operations.[61] Agentic AI can also autonomously enforce these security policies across a diverse range of systems and environments within the organization's IT infrastructure, ensuring consistent application and adherence to security standards.[61] This capability includes the ability to implement and adapt fine-grained access controls, which govern who or what can access specific resources, based on continuous analysis of user behavior, observed network activity, and up-to-the-minute threat intelligence data.[61] By enabling a more dynamic and responsive approach to security, agentic AI allows organizations to adapt their defenses more effectively to the ever-changing threat landscape compared to traditional, static security controls that may become outdated quickly. This adaptability ensures that security measures remain relevant and effective in mitigating emerging risks and protecting sensitive data and critical systems. The ability to dynamically adjust security controls and policies based on real-time conditions represents a significant advancement in strengthening an organization's overall security resilience.

7.4. Potential for Proactive Threat Hunting and Risk Mitigation:

Agentic AI holds substantial potential for enhancing proactive threat hunting activities and improving overall risk mitigation strategies within cybersecurity.[61] These autonomous systems can continuously scan an organization's networks and connected systems to actively search for subtle indicators of compromise (IOCs) and identify potential vulnerabilities that might exist within the infrastructure, often before these weaknesses can be actively exploited by malicious actors.[61] By performing this continuous and automated threat hunting, agentic AI can uncover hidden threats and potential security gaps that might otherwise go unnoticed by traditional security monitoring tools and human analysts. Furthermore, agentic AI can play a crucial role in identifying and prioritizing potential risks based on its continuous analysis of both internal security posture data and external threat intelligence feeds.[61] Once these risks are identified and prioritized, agentic AI can even proactively take steps to mitigate them, such as automatically isolating systems that are deemed to be particularly vulnerable or implementing additional security controls to reduce the likelihood of a successful attack.[61] This capability to proactively hunt for threats and mitigate potential risks represents a significant shift from a reactive security model to a more anticipatory and preventative approach. By autonomously identifying and addressing vulnerabilities before they can be exploited, agentic AI can significantly strengthen an organization's overall security resilience and reduce the likelihood of experiencing damaging cyberattacks.

8. Torq.io's Perspectives and Solutions

8.1. Analysis of Torq.io's viewpoint on security automation and the evolution beyond traditional SOAR:

Torq.io advocates for a paradigm shift in security operations, asserting that the future lies in Autonomous Security Operations achieved through what they term security hyperautomation.[92] This approach, according to Torq.io, represents a necessary evolution beyond the limitations of traditional Security Orchestration, Automation and Response (SOAR) solutions, which they believe have become increasingly obsolete in the face of modern cybersecurity challenges.[92] Their core argument is that the ever-increasing volume and sophistication of cyber threats demand a more intelligent and autonomous approach to security, one that minimizes the reliance on manual intervention and maximizes the efficiency of security operations through the strategic application of artificial intelligence.[92] Torq.io's perspective suggests a fundamental dissatisfaction with the capacity of traditional SOAR platforms to effectively address the complexities of today's threat landscape, particularly in areas such as alert fatigue, the prevalence of false positives, and the resource constraints faced by security teams. They champion the adoption of AI-driven hyper automation as the key to creating a truly autonomous Security Operations Center (SOC) capable of handling the entire threat lifecycle, from initial detection to comprehensive response and remediation, with minimal need for direct human involvement.[92] This viewpoint underscores a belief that only by embracing the power of AI can organizations hope to effectively defend against the rapidly evolving and increasingly sophisticated tactics employed by cyber adversaries.

8.2. Examination of their AI-driven solutions and their approach to autonomous security operations:

Torq.io has developed a suite of AI-driven solutions designed to enable their vision of autonomous security operations. At the core of their approach is a Multi-Agent System comprising various AI agents that are specifically engineered to collaborate and enhance different critical aspects of security operations within a SOC.[92] Torq.io emphasizes the significant benefits that AI brings to security, including the ability to achieve much faster threat detection by analyzing, correlating, and enriching vast quantities of unprocessed security events at machine speed, thereby more effectively identifying genuine threats amidst a sea of alerts.[92] They also highlight the role of AI in enabling faster case prioritization through intelligent case investigation and automated summarization, which allows security analysts to focus their attention and resources on the security incidents that pose the most significant risk and potential impact to the organization.[92] Furthermore, Torq.io introduces Socrates, which they describe as a natural language-driven Agentic AI, specifically designed for the autonomous remediation of critical security threats. This capability aims to dramatically accelerate the mean time to response (MTTR) by allowing the AI to take immediate and decisive actions to neutralize identified threats without requiring manual intervention.[92] A key aspect of Torq.io's approach is their emphasis on the use of natural language prompts to simplify the creation and subsequent deployment of security automations. This user-friendly approach allows security teams to rapidly generate integrations with a wide array of security vendors and to automate complex workflows across these diverse tools without needing extensive coding knowledge or specialized programming skills.[92] Torq.io's AI-driven solutions and their focus on natural language automation suggest a strategic aim to make advanced security automation more accessible and efficient for security teams, ultimately reducing the operational burden on security analysts and significantly accelerating the organization's ability to detect, prioritize, and respond to cyber threats in an increasingly automated and intelligent manner.

8.3. Integration of insights from Torq.io's website (https://torq.io/) throughout relevant sections:

Throughout this report, Torq.io's perspective has been integrated to provide a real-world example of a company advocating for and providing solutions in the realm of AI-enhanced security automation. In the introduction, Torq.io's view on the evolution of SAO towards autonomous security operations driven by AI was mentioned. In the section discussing the challenges of adopting SAO, their argument that traditional SOAR is becoming obsolete will be relevant. The applications of generative and agentic AI sections will benefit from incorporating Torq.io's emphasis on AI-driven threat detection, prioritization, and response as practical examples. When discussing the future impact of AI on SAO, Torq.io's concept of an Autonomous SOC and their innovative use of AI agents will be highlighted as a forward-looking vision for the field. Finally, the conclusion of the report will reference Torq.io's "AI or Die" manifesto to underscore the perceived urgency and critical importance of organizations embracing AI within their security operations to effectively counter the evolving threat landscape. By weaving in Torq.io's specific viewpoints and solutions, this report aims to provide a more concrete understanding of how AI is currently being conceptualized and implemented within the domain of security automation and orchestration.

9. The Future Impact of Generative and Agentic AI on Security Automation and Orchestration

9.1. Synergistic Potential of AI and SAO for Enhanced Cybersecurity Posture:

The integration of generative and agentic AI into Security Automation and Orchestration (SAO) frameworks holds immense synergistic potential for significantly enhancing an organization's overall cybersecurity posture. Generative AI can augment existing SAO capabilities by providing advanced and automated analysis of threat intelligence data, enabling security teams to stay ahead of emerging threats.[40] It can also streamline the creation of essential security content, such as detailed incident reports, comprehensive security policies, and engaging training materials, thereby improving efficiency and ensuring consistency.[40] Furthermore, generative AI's ability to quickly summarize complex security incidents from diverse data sources can significantly accelerate the initial stages of incident response.[40] Agentic AI takes this synergy a step further by enabling truly autonomous threat detection capabilities, allowing organizations to identify and respond to malicious activity in real-time without constant human intervention.[61] Its capacity for intelligent incident response actions, including automated containment and remediation, can drastically reduce the impact of security breaches.[60] Moreover, agentic AI can facilitate the implementation of adaptive security controls and the dynamic enforcement of security policies, allowing organizations to respond more effectively to the evolving threat landscape.[61] The combined power of generative and agentic AI within SAO frameworks promises a significant leap towards a more proactive, highly efficient, and ultimately more resilient cybersecurity posture, reducing the traditional reliance on purely manual processes and leading to demonstrably improved security outcomes for organizations of all sizes.

9.2. Addressing the Challenges and Risks Associated with AI Integration in Security Operations:

While the integration of AI into SAO offers substantial benefits, organizations must also be prepared to address the inherent challenges and potential risks associated with this technological convergence. One significant hurdle involves the integration complexities of AI-powered tools and platforms with existing SAO systems and the broader security infrastructure.[1] Ensuring seamless data flow and interoperability between diverse systems will require careful planning and potentially specialized technical expertise. Furthermore, the development, implementation, and ongoing management of AI-powered SAO solutions will necessitate a skilled workforce with expertise in both cybersecurity and artificial intelligence.[4] Organizations may need to invest in training existing security personnel or recruit individuals with the specific skill sets required to effectively leverage these advanced technologies. Ethical considerations surrounding the use of AI in security operations, such as potential biases embedded within AI models and the risk of unintended consequences from autonomous actions, must also be carefully considered and mitigated through robust governance frameworks.[46] Additionally, the potential for adversarial attacks specifically targeting AI systems in security, such as prompt injection or data poisoning, requires the implementation of appropriate security measures to protect the integrity and reliability of these critical systems.[40] Finally, maintaining an appropriate level of human oversight and control over autonomous AI agents, particularly when making critical security decisions that could have significant operational or financial impacts, will be essential to ensure accountability and prevent unintended negative outcomes.[46] By proactively addressing these challenges and potential risks, organizations can ensure a more secure and effective integration of AI into their security automation and orchestration strategies.

9.3. The Evolving Role of Security Professionals in an AI-Augmented SOC:

The increasing integration of generative and agentic AI into Security Operations Centers (SOCs) will inevitably lead to an evolution in the roles and responsibilities of security professionals. As AI systems become more adept at handling routine and repetitive tasks, such as initial alert triage, basic incident investigation, and the generation of standard reports, security analysts will likely see a shift in their focus towards more complex and strategic activities.[40] This could involve dedicating more time to in-depth investigations of sophisticated cyberattacks, developing and refining incident response strategies, conducting advanced threat hunting exercises to proactively uncover hidden threats, and taking a more strategic approach to overall security planning and architecture.[40]

Furthermore, the rise of AI in security is likely to create new specialized roles within security operations teams. These roles might focus on the development, training, and governance of the AI models that underpin these advanced security solutions, ensuring their effectiveness, accuracy, and ethical use.[21] To remain relevant and effective in this evolving landscape, continuous learning and upskilling will be paramount for security professionals. They will need to develop a deeper understanding of AI technologies, including how they work, their potential capabilities within cybersecurity, and the best practices for their secure and responsible implementation.[21] While AI will undoubtedly automate many of the more mundane and repetitive aspects of security operations, it is unlikely to completely replace the need for human expertise, critical thinking, and nuanced judgment. Instead, the future of cybersecurity in an AI-augmented SOC will likely involve a collaborative partnership between human professionals and intelligent AI systems, where each leverages their respective strengths to create a more robust and resilient security posture.

10. Conclusion

Generative and agentic AI are poised to revolutionize the landscape of security automation and orchestration, offering transformative potential for enhancing cybersecurity defenses. The ability of generative AI to analyze vast datasets, generate insightful content, and streamline reporting, coupled with the autonomy and decision-making capabilities of agentic AI for real-time threat detection and response, represents a significant leap forward in our ability to manage and mitigate cyber risks.

These technologies promise improved efficiency, faster incident response times, enhanced accuracy, and greater scalability for security operations. However, the integration of AI into SAO is not without its challenges. Organizations must carefully navigate integration complexities, address the need for skilled personnel, manage expectations to avoid over-automation, and ensure continuous monitoring and adaptation of their AI-powered security systems. The role of security professionals will also evolve, shifting towards more strategic and complex tasks, requiring continuous learning and adaptation to this rapidly changing technological landscape. Ultimately, the synergy between AI and SAO holds the key to building a more proactive, efficient, and resilient cybersecurity posture. As Torq.io aptly suggests with their "AI or Die" manifesto, embracing the power of artificial intelligence in security operations is becoming increasingly critical for organizations to effectively defend against the ever-growing sophistication and volume of cyber threats in the digital age.

Feature	Traditional SOAR	AI-Powered Autonomous Security Operations (as advocated by Torq.io)
Scope of Automation	Primarily focused on predefined workflows and tasks	Handles entire threat lifecycle from detection to response
Decision-Making Capability	Relies on pre-configured rules and human input	Autonomous decision-making based on AI analysis
Human Intervention	Significant human oversight and intervention required	Minimal human intervention, with AI handling most tasks

Adaptability	Limited adaptability to new or unknown threats	High adaptability through continuous AI learning
Threat Coverage	Effective against known threats and scenarios	Aims to detect and respond to both known and novel threats
Analyst Workload	Can reduce repetitive tasks but still high	Significantly reduced through AI-driven automation
MTTD/MTTR	Improvement over manual processes	Drastically reduced through machine-speed analysis and autonomous response

Application Area	Generative AI Role	Agentic AI Role
Threat Intelligence	Automated Analysis, Report Generation	Autonomous Detection of Anomalies
Security Content	Training Material Creation, Policy Generation, Code Generation, Report Generation	N/A
Incident Response	Incident Summarization, Report Generation, Suggesting Remediation Actions	Intelligent Response Actions (Containment, Remediation)

Threat Detection	Pattern Recognition, Anomaly Detection	Autonomous Threat Detection, Behavioral Analysis
Vulnerability Management	Predictive Analysis, Patch Recommendation	Proactive Threat Hunting, Risk Prioritization
Security Controls	Generating Configuration Scripts	Adaptive Security Controls, Autonomous Policy Enforcement

Works cited

Can be found at https://www.c-ooda.com/book-of-five-keys

Appendix D: Disrupting Attacker Operations in Cyberspace

Ethical and Legal Considerations

I. Introduction: The Concept of Disrupting Attacker Operations in Cyberspace

The notion of "taking the fight to the enemy" in the context of cybersecurity signifies a paradigm shift from purely reactive defense to more proactive strategies aimed at countering and neutralizing threats at their source or during their operational stages. This concept encompasses a range of activities, from deploying sophisticated detection mechanisms that actively engage with adversaries within a defended network to undertaking measures intended to impair or halt ongoing malicious campaigns originating from external sources. The increasing frequency and sophistication of cyberattacks targeting individuals, organizations, and critical infrastructure have underscored the limitations of traditional, passive security postures.[1] Consequently, the exploration of methods to disrupt attacker operations has gained significant traction within the cybersecurity community and among policymakers.

However, this proactive stance introduces a complex web of ethical and legal considerations that must be carefully navigated. The global and interconnected nature of cyberspace transcends traditional territorial boundaries, complicating the application of established legal frameworks and raising novel ethical dilemmas. Determining the permissibility and scope of actions intended to disrupt attackers requires a thorough understanding of international law, domestic legislation, and evolving ethical norms in the digital realm.

This report will delve into these multifaceted considerations, examining the ethical implications of offensive cyber operations and disruption activities, analyzing the relevant international legal framework, exploring the concept of active defense, presenting expert analyses on the risks and benefits involved, investigating pertinent case studies, examining the United States' legal framework in this domain, discussing the potential for escalation and unintended consequences, and analyzing the principle of proportionality in the context of disrupting attacker operations.

II. Ethical Implications of Offensive Cyber Operations and Disruption

The ethical landscape of offensive cyber operations and the disruption of attacker activities is intricate, shaped by various philosophical paradigms that offer differing perspectives on the permissibility and scope of such actions.[3] For instance, Kantian ethics, emphasizing the inherent dignity of persons, would likely scrutinize any cyber operation that could potentially harm innocent individuals or use them as mere means to an end.[3] Conversely, utilitarian perspectives might weigh the potential benefits of disrupting an attack, such as preventing widespread harm, against the potential negative consequences of the disruptive action itself.[4] Virtue ethics, focusing on the character and integrity of the decision-maker, would emphasize the importance of honesty, integrity, and prudence in the deployment of such capabilities.[4] Furthermore, cultural and societal values also play a crucial role in shaping ethical justifications for actions in cyberspace.[3] The absence of a universally accepted ethical framework necessitates a comprehensive understanding of how diverse moral traditions inform the justification for cyber actions, highlighting the challenge in establishing globally coherent norms.[3]

Traditional Just War Theory, with its focus on physical harm and tangible damage, encounters difficulties when applied to the domain of cyberwarfare, where significant harm can be inflicted without causing direct injury or death.[5] This raises fundamental questions about when and how offensive cyber actions can be ethically justified under existing moral frameworks that were primarily developed in the context of kinetic conflict.[5] The ethical responsibilities of computing professionals, as outlined by organizations like the ACM, provide a foundational set of principles that emphasize benefiting society, avoiding harm, respecting privacy, and honoring confidentiality.[6] Disrupting attacker operations can be viewed as aligning with the ethical duty to protect data and systems from threats.[7] However, the proactive nature of such measures introduces a tension with the principle of avoiding negative consequences and respecting legal boundaries.[6] The ethical obligation to defend against cyber threats might indeed justify proactive measures against attackers, but the methods employed must still adhere to broader ethical principles such as proportionality and minimizing harm to non-targets.[7]

The concept of "hacking back," or retaliating against attackers by infiltrating their systems, presents a complex ethical dilemma.[1] Proponents argue that it can serve as a deterrent to cybercriminals and provide organizations with an advantage in gathering intelligence and disrupting ongoing attacks.[1] However, opponents raise significant ethical concerns about vigilantism, the undermining of the rule of law, the risk of escalating conflicts in cyberspace, and the potential for collateral damage to innocent parties.[1] The inherent difficulty in accurately attributing cyberattacks further exacerbates these ethical concerns, as retaliating against the wrong entity can have severe and unjust consequences.[1]

Active defense, which involves proactively detecting, disrupting, and countering adversaries within one's own network, also lies at the intersection of defense and aggression, raising ethical questions about the limits of protection versus provocation.[4] While the intent behind active defense is typically protective, tactics such as deception and the use of honeypots require careful ethical evaluation regarding their potential consequences and the risk of escalation.[4] The ethical nature of these tactics often hinges on intent, whether the primary goal is to protect or to cause harm.[4]

To mitigate the ethical risks associated with offensive cybersecurity expertise, establishing ethical principles within training programs is crucial.[8] Principles such as proportionality and necessity, respect for privacy and confidentiality, avoiding harm, transparency and disclosure (within the training context), and accountability and responsibility aim to guide future practitioners in the ethical application of their skills.[8]

The increasing integration of artificial intelligence into cybersecurity introduces new ethical challenges.[9] These include concerns about bias and fairness in AI algorithms, the lack of transparency in some AI decision-making processes, and questions of accountability when AI systems make mistakes.[9] As AI is increasingly used in both offensive and defensive cyber operations, these ethical dilemmas require careful consideration to ensure responsible and justifiable actions in cyberspace.[9]

III. International Legal Framework for Cyber Warfare and Intervention

The international legal landscape governing cyber warfare and intervention is complex and still evolving. While there is a general consensus that existing international law applies to state activities in cyberspace, the precise manner of its application remains a subject of ongoing debate and clarification.[10] Key legal instruments such as the UN Charter, the Budapest Convention, and the ongoing efforts to establish a new UN cybercrime treaty form the basis of this framework.[12]

The UN Charter, drafted in the pre-cyber era, establishes fundamental principles relevant to state conduct in cyberspace, including the sovereignty of states, the prohibition of the use of force, the principle of non-intervention in the internal affairs of other states, the encouragement of peaceful settlement of disputes, and the inherent right of individual or collective self-defense if an armed attack occurs.[14] However, the application of these principles to cyber operations presents significant challenges, particularly in defining what constitutes a "use of force" or an "armed attack" in the digital realm.[14]

Several international treaties address cybercrime from a law enforcement perspective. The Palermo Convention aims to combat transnational organized crime, and while not specific to cyber activities, its provisions on cooperation and extradition are relevant.[12] The Budapest Convention on Cybercrime is the first international treaty specifically aimed at reducing computer-related crime by harmonizing national laws, improving investigative techniques, and increasing international cooperation.[12] The ongoing negotiations for a new UN cybercrime treaty reflect a global effort to enhance international cooperation in preventing and investigating cybercrime, although disagreements persist regarding its scope and human rights safeguards.[13] These treaties primarily focus on criminal offenses and do not directly address state-sponsored offensive cyber operations or the legality of disrupting attacker operations in a national security context.[12]

The principle of state sovereignty is a cornerstone of international law, granting each state supreme authority over its territory, including the cyber infrastructure and activities within its borders.[14] While there is agreement that sovereignty applies in cyberspace, the threshold for what constitutes a violation, particularly in the absence of physical damage or loss of functionality, remains contested.[18] Some legal experts view any unauthorized intrusion into another state's cyber infrastructure as a violation of sovereignty, while others argue that a certain level of effect or interference with governmental functions is necessary to cross this threshold.[19] The principle of non-intervention prohibits coercive interference by one state in the internal or external affairs of another state.[19] In the cyber context, this principle would prohibit coercive cyber operations aimed at influencing a state's sovereign decisions, such as interfering with elections or essential governmental services.[21]

The legality of intervening in another nation's cyber infrastructure for defensive purposes is closely tied to the right of self-defense under Article 51 of the UN Charter.[14] This right is triggered if an "armed attack" occurs. The critical question is whether a cyberattack can reach this threshold. While a destructive cyberattack causing significant physical damage or loss of life would likely be considered an armed attack, many cyber operations, such as espionage or disruption of non-essential services, do not meet this high bar.[14] Even for cyber activities below the threshold of an armed attack, the principles of sovereignty and non-intervention still apply, requiring careful consideration of the legality of any interventionist measures.[22] The difficulty of attributing cyberattacks to specific state actors further complicates the legal justification for defensive interventions.[15]

IV. Active Defense in Cybersecurity: Ethical and Legal Boundaries

Active defense in cybersecurity represents a proactive approach to protecting networks and systems by actively detecting, disrupting, and countering cyber threats.[23] It encompasses a range of techniques aimed at increasing the cost and complexity for attackers, primarily focusing on actions taken within the defender's own network.[24] Common active defense methods include the deployment of honeypots, which are decoy systems designed to attract and trap attackers; deception technologies, which involve creating fake assets and information to mislead adversaries; threat intelligence gathering and analysis to anticipate and identify attacks; active monitoring of network traffic and system behavior for anomalies; and threat hunting, which involves proactively searching for signs of malicious activity.[23] Automated incident response systems also play a crucial role in active defense by rapidly taking action to contain and mitigate detected threats.[26]

While active defense primarily focuses on defensive measures within an organization's own infrastructure, the term can sometimes encompass offensive cyber operations, particularly in military and state contexts.[28] For instance, NATO's definition of active defense includes preemptive or counter-operations against the source of an attack.[29] China's military strategy also employs the concept of "active defense".[28] This broader interpretation in state-level contexts can blur the lines between purely defensive actions and offensive operations aimed at neutralizing threats before or during an attack.[29]

The legal boundaries of proactive cyber defense for private entities are a subject of ongoing debate.[30] While traditional proactive measures focused on securing one's own network perimeter are generally accepted as legal and ethical,[32] actions that extend beyond these boundaries, such as "hacking back" into attacker systems, are generally prohibited by laws like the Computer Fraud and Abuse Act (CFAA) in the United States.[30]

Proposals like the Active Cyber Defense Certainty Act (ACDCA) have aimed to amend the CFAA to allow certain forms of hack-back under specific conditions, primarily for attribution, disruption, or monitoring malicious activity.[31] However, concerns about international security, inter-state relations, and the potential for abuse have hindered the widespread legalization of such measures for private actors.[30]

Active defense tactics raise specific ethical considerations.[4] The use of deception, while intended for protective purposes, can be ethically ambiguous. Questions arise regarding the transparency of such tactics and the potential for unintended consequences or harm to non-targets.[4] Establishing clear rules of engagement and ensuring accountability for the deployment of active defense measures are crucial for addressing these ethical concerns.[4]

V. Risks and Benefits of Disrupting Attacker Operations: Expert Analysis

Cybersecurity experts and organizations widely acknowledge the increasing sophistication and frequency of cyber threats, including malware, ransomware, phishing attacks, and supply chain compromises, which can inflict significant financial, operational, and reputational damage on organizations.[2] In this context, the ability to disrupt attacker operations offers potential benefits such as deterring future attacks by imposing costs on adversaries and gathering valuable intelligence about their tactics, techniques, and infrastructure.[1] Proactive measures can also provide defenders with a crucial advantage in understanding and mitigating threats before they can achieve their objectives.[26]

However, disrupting attacker operations is not without significant risks. A primary concern is the potential for escalation of cyber conflicts.[1] Offensive actions, whether by states or private entities, can provoke retaliation from adversaries, leading to a cycle of attacks and counterattacks that can rapidly spiral out of control.[15] The challenge of accurate attribution in cyberspace further exacerbates this risk, as misidentifying the attacker can lead to escalatory actions against the wrong target.[1] Moreover, the interconnected nature of cyberspace increases the risk of unintended consequences, where disruptive actions aimed at attackers could inadvertently affect innocent third parties, critical infrastructure, or even the broader internet ecosystem.[40]

Experts in state-level cyber operations advocate for a cautious and strategic approach to offensive capabilities, emphasizing the need for increased transparency, a nuanced understanding of their utility, and robust risk mitigation measures.[18] There is a general agreement that offensive cyber operations should not be viewed as a panacea and should be employed judiciously, with clear authorization and oversight at the highest levels of government.[42] Some experts even argue that prioritizing offensive operations can be counterproductive, increasing national vulnerabilities and international tensions, and suggest that a stronger focus on defensive capabilities might be a more effective and less destabilizing approach.[44]

VI. Case Studies of Disrupted Attacker Operations

Publicly available examples of successfully disrupted attacker operations are often limited due to the sensitive nature of such activities and concerns about revealing capabilities and tactics. However, numerous instances of significant cyber operations and attacks have been publicly attributed, providing insights into the types of threats that defenders might seek to disrupt.[45] These include state-sponsored campaigns like Moonlight Maze, Titan Rain, and Operation Buckshot Yankee, which involved espionage and network penetration.[46] The Stuxnet attack against Iran's nuclear program is a notable example of a sophisticated cyber operation aimed at disrupting critical infrastructure.[47] More recent examples include cyberattacks attributed to Russia, China, Iran, and North Korea, targeting various sectors for espionage, disruption, and financial gain.[48]

The increasing targeting of specific sectors, such as law firms holding sensitive client data[49] and the healthcare industry,[50] underscores the high stakes of cyberattacks and the critical need for effective defense mechanisms. While these examples primarily illustrate successful attacks, they highlight the types of operations that defenders aim to prevent or disrupt.

The fictional case study of "The Great Cyberwar of 2002" provides a framework for exploring the complex legal and ethical questions that arise in the context of large-scale cyber conflict, including the definition of "armed attack" and the status of cyberwarriors.[51] While not a real-world example of disruption, it helps to frame the debates surrounding potential responses to significant cyber incidents.

The four phases of cyber warfare escalation observed in modern conflicts, such as the Israel-Hamas conflict and the Russia-Ukraine War, illustrate the dynamic nature of cyberattacks and potential opportunities for disruption at various stages of escalation.[52]

Analyzing these instances reveals that the ethical and legal debates surrounding disrupted attacker operations often center on the challenges of attribution, the proportionality of responses, and the potential for escalation or unintended consequences.[1] The lack of clear international norms and the differing legal frameworks across nations further complicate these debates.[1]

VII. Legal Framework within the United States

The United States has established a legal framework concerning actions to disrupt cyberattacks originating from outside the country, primarily through Title 10 of the U.S. Code, which grants authority to the Secretary of Defense to conduct military cyber activities or operations, including clandestine ones, to defend the nation.[54] This authority extends to operations short of hostilities for purposes such as preparation of the environment, information operations, force protection, deterrence, and counterterrorism, and explicitly affirms the right to respond to malicious cyber activity by foreign powers.[54]

The U.S. Cyber Command (USCYBERCOM) serves as the key military entity responsible for directing and coordinating cyberspace operations to defend the Department of Defense information networks and the nation from significant cyberattacks.[55] Operating globally, USCYBERCOM has evolved to incorporate offensive capabilities as part of its "defend forward" strategy, which involves proactively engaging adversaries to disrupt or halt malicious cyber activity at its source.[55]

The Computer Fraud and Abuse Act (CFAA) (18 U.S.C. § 1030) is the primary federal law for prosecuting cybercrime in the United States, including unauthorized access to and damage of computers used in interstate or foreign commerce.[34] The CFAA has extraterritorial application, allowing for the prosecution of individuals located outside the U.S. who target American computer systems.[59] While the CFAA provides a legal tool for addressing cyberattacks originating from abroad, its interpretation, particularly concerning actions that might be considered "hacking back" by private entities, remains complex and subject to legal debate.[60]

Beyond these specific legal instruments, the U.S. has a broader legal and regulatory landscape focused on improving overall cybersecurity posture, managing risks, and responding to incidents, including data breach notification laws and the role of agencies like the Securities and Exchange Commission (SEC) in cybersecurity disclosures.[61]

Section 167b of Title 10 further outlines the authority and mission of US Cyber Command, emphasizing its role in defending national interests through collaboration with domestic and international partners.[64] This reinforces the legal mandate and collaborative nature of the U.S.'s approach to cyberspace defense.

VIII. Escalation and Unintended Consequences

A significant concern associated with disrupting attacker operations, especially through offensive measures, is the potential for escalation of cyber conflicts.[39] Retaliatory actions can easily trigger a cycle of attacks and counterattacks, leading to an unpredictable and potentially damaging spiral.[1] The inherent difficulties in accurately attributing cyberattacks increase the risk of escalation against the wrong adversary.[15] Furthermore, the complex and interconnected nature of cyberspace means that disruptive actions can have unintended consequences, such as affecting innocent third parties, damaging critical infrastructure, or destabilizing the broader internet environment.[40]

Understanding common attacker tactics, such as privilege escalation, is crucial for developing effective defensive strategies, including disruption, while minimizing the risk of unintended effects.[66] The phases of cyber warfare escalation observed in recent conflicts provide a framework for comprehending how cyber incidents can intensify and potentially coordinate with kinetic operations, underscoring the high stakes involved.[52]

Even actions intended to deter cyberattacks carry inherent risks of unintended consequences and escalation.[69] The strategic risks of offensive cyber operations include undermining the security of the internet and acting as an escalatory trigger, necessitating precise execution, robust control, and continuous monitoring of their effects.[41]

Cyber retaliation, in particular, carries substantial risks, including misattribution, disproportionate responses, escalation, and the potential to undermine international law and norms.[15] Its effectiveness as a deterrent is also questionable, as it might provoke further attacks.[15] The significant potential damage from cyberattacks underscores the importance of careful and measured responses, as ill-conceived retaliation could exacerbate the harm.[72]

IX. Proportionality in Disrupting Attacker Operations

The principle of proportionality in international law, especially within the context of armed conflict (IHL), prohibits attacks where the expected incidental civilian harm is excessive in relation to the concrete and direct military advantage anticipated.[74] This principle applies to cyber operations, requiring a careful balancing of the anticipated military advantage of disruptive actions against the potential harm to civilians and civilian infrastructure, considering both immediate and cascading effects.[75] The assessment of proportionality is made ex ante based on the information reasonably available at the time of the operation.[74]

Applying the principle of proportionality to cyberattacks presents several challenges and ambiguities.[78] The potential for significant civilian loss without direct physical harm and the difficulties in accurately assessing the long-term and indirect consequences of cyber operations complicate the application of this principle.[78] There is a recognized need for clearer proportionality standards specifically tailored to the cyber domain.[78]

Any disruptive actions taken as countermeasures to cyberattacks must also adhere to the principle of proportionality, ensuring that the response is not excessive in relation to the initial injury and is aimed at legitimate defensive and deterrent purposes.[79] Countermeasures should be directed only at the responsible state and must satisfy the principles of necessity and proportionality, ceasing once the original violation ends.[79]

The concept of collateral damage, traditionally defined as unintended harm to civilians or civilian objects, also applies to cyber operations.[83] While traditional definitions focus on physical harm, cyber operations can cause significant harm to data, systems, and critical infrastructure, potentially leading to real-world consequences for civilians.[83] The interconnected nature of cyberspace makes it challenging to precisely target cyber operations and contain their effects, increasing the risk of collateral damage to civilian systems and populations.[87] Therefore, any disruptive actions must carefully consider the potential for harm to innocent parties and ensure that the military advantage anticipated outweighs the likely collateral damage.[82]

X. Conclusion: Navigating the Complexities of Disrupting Attacker Operations

Disrupting attacker operations in cyberspace presents a compelling strategy for enhancing cybersecurity, moving beyond traditional reactive measures to actively counter and neutralize threats. However, this proactive approach is fraught with ethical and legal complexities that demand careful consideration. Ethically, the deployment of offensive cyber capabilities and disruptive tactics must be weighed against principles of proportionality, the potential for harm to innocent parties, and the risk of escalating conflicts. The lack of a universally accepted ethical framework for cyber operations necessitates a nuanced understanding of diverse moral perspectives and professional responsibilities.

Legally, the international framework, while applicable to cyberspace, requires further clarification and adaptation to the unique characteristics of the digital domain. Principles of state sovereignty, non-intervention, and the right to self-defense provide a foundation, but their interpretation and application to cyber activities remain contested. The ongoing development of international agreements and the evolution of state practice will continue to shape the legal landscape. Within the United States, a legal framework exists for military cyber operations aimed at defending national interests, but the legal boundaries for private sector actions to disrupt attackers are still under debate.

The risks of escalation and unintended consequences associated with offensive cyber actions and retaliation underscore the need for caution, precision, and robust oversight. The principle of proportionality serves as a critical constraint, requiring a careful balance between the anticipated military advantage and the potential harm to civilians and civilian infrastructure.

Navigating these complexities requires a multifaceted approach that integrates ethical considerations, adherence to legal frameworks, a thorough understanding of the risks and benefits involved, and ongoing dialogue among experts, policymakers, and the international community. As cyberspace continues to evolve as a domain of conflict and competition, the responsible and lawful disruption of attacker operations will remain a critical challenge requiring careful deliberation and a commitment to minimizing harm while safeguarding national security and global stability.

Key Valuable Tables:

1. **Ethical Paradigms and Their Application to Offensive Cyber Operations:**

Ethical Paradigm	Core Principles	Potential Justifications for Offensive Cyber Operations	Potential Limitations on Offensive Cyber Operations
Kantianism	Human dignity, persons as ends	Protecting critical infrastructure from attacks that could harm individuals.	Avoiding any action that could use innocent individuals as a means to an end or cause indiscriminate harm.
Utilitarianism	Greatest good for the greatest number	Disrupting attacks that could cause widespread harm or significant disruption to society.	Ensuring that the benefits of disruption outweigh the potential harm caused by the action itself, including to the attacker and third

			parties.
Virtue Ethics	Honesty, integrity, prudence	Acting with integrity and prudence to defend against malicious actors.	Avoiding deceptive or reckless actions that could have unintended negative consequences.
Confucianism	Benevolent rule, peaceful order	Maintaining a peaceful and secure cyberspace for the benefit of the populace.	Ensuring that disruptive actions are carried out by legitimate authorities and contribute to overall stability.

2. **Key International Treaties Relevant to Cyber Activities:**

Treaty Name	Key Provisions Relevant to Cyberspace	Focus	Limitations or Challenges in Application

UN Charter	Sovereignty, prohibition of use of force, non-intervention, right to self-defense.	International Law, Use of Force	Difficulty in defining "use of force" and "armed attack" in cyberspace; attribution challenges.
Budapest Convention on Cybercrime	Criminalizes various cyber offenses, promotes international cooperation in investigations.	Cybercrime, International Cooperation	Primarily focused on criminal activity, less direct application to state-sponsored offensive operations.
Second Additional Protocol to the Convention on Cybercrime	Enhances cooperation and disclosure of electronic evidence.	Cybercrime, Evidence Sharing	Not yet in force as of December 2022.
(Proposed) UN Cybercrime	Aims to criminalize cyber-dependent	Cybercrime, International Cooperation	Ongoing negotiations, disagreements on scope

Treaty	and cyber-enabled crimes, enhance international cooperation.		and human rights safeguards.

3. Comparison of Active Defense Techniques:

Technique	Description	Potential Benefits in Disrupting Attackers	Ethical and Legal Considerations
Honeypots	Decoy systems designed to attract and trap attackers.	Gathers intelligence on attacker tactics, can divert attackers from real systems.	Deception may raise ethical concerns; need to ensure no unintended harm to third parties.
Deception Technologies	Creating fake assets, data, or credentials	Detects attacker presence,	Deception may raise ethical

	to mislead attackers.	tracks movements , wastes attacker resources.	concerns; need to ensure no unintended harm to third parties.
Threat Intelligence	Gathering and analyzing information about potential threats.	Proactive detection and response, better understanding of attacker TTPs.	Ethical collection and use of intelligence data.
Active Monitoring	Continuous surveillance of networks and systems for suspicious activity.	Early detection of breaches and malicious behavior.	Privacy concerns related to monitoring user activity.
Threat Hunting	Proactive searching for hidden malicious activity within a	Identifies threats that may evade traditional security measures.	Requires skilled personnel and deep understand

	network.		ing of network and threat landscape.
Automated Incident Response	Systems that automatically take action to isolate, block, or mitigate threats.	Rapid response to known threats, reduces dwell time for attackers.	Risk of false positives and unintended disruptions to legitimate activity.

4. Summary of US Legal Framework for Disrupting Cyberattacks:

Law/Author ity	Key Provisions/ Mandates	Relevance to Disrupting Attacker Operations	Limitation s or Areas of Debate
10 U.S. Code § 394	Secretary of Defense authorized to conduct military	Provides legal basis for proactive defense against	Scope of "defense" and "preparatio

	cyber operations, including clandestine ones, to defend the U.S.	foreign cyber threats.	n of the environme nt" can be broad; oversight mechanism s.
US Cyber Command USCYBERCOM	Directs and coordinates military cyberspace operations, defends DoD and the nation from cyberattacks.	Operational arm for executing defensive and offensive cyber operations under Title 10 authority; "defend forward" strategy.	Balance between defensive and offensive roles; coordinatio n with other agencies.
Computer Fraud and Abuse Act (CFAA) (18 U.S.C. § 1030)	Prohibits unauthorized access to and damage of computers used in interstate or foreign commerce; has extraterritori	Provides legal tool for prosecuting cyberattacks originating from outside the U.S.	Interpretati on of "unauthori zed access" debated; restrictions on private sector "hacking

	al application.		back."
§167b of Title 10	Outlines authority and mission of US Cyber Command, emphasizing collaboration.	Reinforces legal mandate for national defense in cyberspace and importance of partnerships.	Subject to ongoing legislative updates and interpretations.

Works cited

Can be found at https://www.c-ooda.com/book-of-five-keys

Appendix E - General Resources

Here is a list of references the reader could use to do further research into the subjects discussed:

General Cybersecurity Resources:

- National Institute of Standards and Technology (NIST): NIST provides a wealth of cybersecurity resources, frameworks, and best practices. Their Cybersecurity Framework is a widely adopted guide.

    o   csrc.nist.gov

- Cybersecurity and Infrastructure Security Agency (CISA): CISA is the U.S. government agency responsible for strengthening national cybersecurity and infrastructure protection. Their website offers alerts, advisories, and guidance.

    o   www.cisa.gov

- SANS Institute: SANS offers a wide range of cybersecurity training courses, certifications, research, and resources.

    o   www.sans.org

- OWASP Foundation: The Open Web Application Security Project (OWASP) is a non-profit organization dedicated to improving the security of software.[1] They provide valuable resources on web application security.

    o   owasp.org

- Information Systems Audit and Control Association (ISACA): ISACA focuses on IT governance, audit, security, and risk management. They offer frameworks and certifications like CISA and CISM.

    o   www.isaca.org

Specific to Phishing and Email Security:

- Anti-Phishing Working Group (APWG): APWG is an industry coalition focused on combating phishing and email fraud. Their website offers reports and resources.
  - apwg.org

- CISA's Stop Think Connect Campaign: Provides resources and tips for individuals and organizations on staying safe online, including information on phishing.
  - www.cisa.gov/news-events/news/stopthinkconnect

Specific to Web Application Attacks and Security:

- OWASP Top Ten: OWASP publishes a regularly updated list of the ten most critical web application security risks.
  - owasp.org/www-project-top-ten/
- PortSwigger Web Security Academy: Offers free online training and resources on web application security vulnerabilities and how to exploit and prevent them.
  - portswigger.net/web-security

Specific to Data Breaches and Data Loss Prevention:

- Verizon Data Breach Investigations Report (DBIR): An annual report providing insights into data breach trends and patterns.
  - Search online for the latest "Verizon Data Breach Investigations Report."
- Privacy Rights Clearinghouse: Offers a comprehensive database of data breaches and resources on data privacy.
  - privacyrights.org
- Information Commissioner's Office (ICO) (UK): Provides guidance on data protection and data breach reporting, which can be relevant globally.
  - ico.org.uk

Specific to Emerging Threats (AI-Powered Attacks, Quantum Computing):

- MIT Technology Review: Often publishes articles on the latest advancements and implications of AI and quantum computing.
  - www.technologyreview.com
- IEEE Spectrum: Provides news and analysis on technology trends, including AI and quantum computing.
  - spectrum.ieee.org
- NIST's Post-Quantum Cryptography Program: NIST is actively working on standardizing post-quantum cryptography algorithms. Their website provides updates and information.
  - csrc.nist.gov/projects/post-quantum-cryptography
- Various Cybersecurity Vendor Blogs and Research Papers: Many cybersecurity companies publish blogs and research papers on emerging threats. Searching for terms like "AI in cybersecurity," "quantum cryptography," and "emerging cyber threats" will yield relevant results from vendors like Microsoft, Google, IBM, and specialized security firms.

For Ongoing Learning:

- Reputable Cybersecurity News Outlets: Stay updated on the latest threats and trends by following cybersecurity news websites like Krebs on Security, The Hacker News, SecurityWeek, and Dark Reading.

This list provides a starting point for further exploration. The cybersecurity landscape is constantly evolving, so it's important to stay informed through a variety of sources. Remember to critically evaluate the information you find and consider the source's credibility.

Appendix F - The Original Blogs (2010)

Note: The original links are included as is, and I have done no checking to see if they are still live or not. This is the blog the way it was posted in 2010.

Welcome to The Way of Cyber Strategy

Welcome to my small corner of the Internet where I hope to share my experiences and lessons learned in the art of cyber security strategy. I have chosen to blog anonymously so I may be free to express opinions that may be contrary to employers and peers in my industry. In my world of information assurance, I have often pondered why there are so many companies and consultants promoting the latest and greatest in compliance, controls and tools, without any thought of using strategic thinking in the cyber battlefield where my kind live. My musings always seem to go back to a simple book written a long time ago, The Book of Five Rings by Miyamoto Musashi where the strategy of warfare is laid out in very simple terms.

Samurai history and thought has always been an interest of mine, as well as strategic thinking. Things like USAF Colonel John Boyd's OODA Loop (for Observe, Orient, Decide and Act) concept have always fascinated me. I have always applied these strategies and concepts in my day to day work, but very rarely see any text or program that teaches one how to integrate them into current day cyber security efforts.

I previously attended an executive-level conference on cyber security, and the lack of speakers and educational tracks with clear strategic value prompted me to actually start my blog!

In subsequent posts I plan on relating how The Book of Five Rings can be used as Five Keys to Successful Cyber Security Strategy. As events allow

I will also post real world examples of the strategy in action (with anonymous sources of course).

Comments, musings and links to other strategic sources are always welcome!

Five Keys to Successful Cyber Security Strategy - Introduction

Google search for the term "cyber security" returns 1.6 million hits. Adding "strategy" to that term reduces the results to around 16,200, with the top result being Obama Administration Outlines Cyber Security Strategy - Security Fix, a Washington Post article. The article is about "Securing Cyberspace for the 44th Presidency, A Report of the CSIS Commission on Cybersecurity for the 44th Presidency'. Interesting reading, but as Brian Krebs reported in the Washington Post on December 8, 2008, "The Obama team said it plans to work with industry and academia to "develop and deploy a new generation of secure hardware and software for our critical cyber infrastructure," and "work with the private sector to establish tough new standards for cyber security and physical resilience." They also pledge to help combat cyber espionage and "initiate a grant and training program to provide federal, state, and local law enforcement agencies the tools they need to detect and prosecute cyber crime."

OK. Lots of good information, ideas, tools and how to, but is a compilation of tactical plans such as hardware, software and standards really a "strategy?" At least that is a question that has crossed my mind on more than one occasion.

A strategic thinker looks at a broader picture, a future state and the goals of your organization. A strategic thinker understands the tools or weapons at his disposal. A strategic thinker knows how to employ tactics in accomplishing his strategy. A successful cyber strategist knows how to engage his enemy in battle. A winning strategist understands the Way of his opponent. A true cyber strategist knows that they do not know everything and are willing to learn.

These traits are more fully explained for the Samurai in his Five Books, or Rings; The Ground Book, the Water Book, the Fire Book, the Wind Book and the Book of the Void. His books correlate with my Book of Five Keys. Those keys are Self Knowledge, Agility, Action, Threats, and Process.

The first Key, Self Knowledge relates to Musashi's Ring, The Ground Book. We will explore that relationship in my next entry.

The Key Of Knowledge (Earth Book)

Musashi identified four Ways in which man passes through life, "as gentlemen, farmers, artisans and merchants." I liken the tools of my trade to weapons to combat those who seek to combat my system, so the Way of the warrior is the view I take of my profession. According to Musashi, "The Way of the warrior is to master the virtue of his weapons. If a gentleman dislikes strategy he will not appreciate the benefit of weaponry, so must he not have a little taste for this?"

Cyber Strategy is similar in that there are multiple "Ways" or disciplines to learn. In that discipline there may be many tools and processes that are useful, and it is the Strategist's quest to learn these tools to the best of his ability.

Musashi's stated meaning in the Earth book is " I give an overall picture of the art of fighting and my own approach. It is difficult to know the true Way through swordsmanship alone. From large places one knows small places, from the shallows one goes to the depths. Because a straight road is made by leveling the earth and hardening it with gravel, I call the first volume Earth, as if it were a straight road mapped out on the ground."

I equate this book to Knowledge, or being aware of your cyberworld. Primarily there are four areas of knowledge

1. Situational Awareness

2. Knowing your environment

3. Your tools or controls

4. Security Awareness

Let's take the first, situational awareness. I like the definition in Wikipedia, " Situational Awareness is the perception of environmental elements within a volume of time and space, the comprehension of their meaning, and the projection of their status in the near future." In the information assurance world our environmental elements can be viewed as elements of the risk equation. What are the threats against our environment? What our vulnerabilities? What assets are we trying to protect?

There are many schools and thoughts about risk management, so I will not delve into those in this blog. If you, as a budding Cyber Strategist, are new to the world of risk management, or still believe that security drives business, concentrate your situational awareness efforts on learning the way of risk management.

In my next entry, we will discuss how a Cyber Strategist looks at knowing the environment.

Recommended web links: Risk Analysis

The Key Of Knowledge (Earth Book) Part 2

The second area of knowledge in this key is "Knowing your environment." There are many aspects to this knowledge, such as, but not limited to asset classification and network topology. In order to look at the security of your environment in a strategic manner, understanding where and what your most critical assets are is paramount to success. Many times when speaking with peers, I find that asset classification has been placed on a back burner, or the importance is lost in the pressures of fighting fires and politics. And perhaps the fact that a decent classification policy and process can be difficult to develop, implement and maintain. It is imperative you take this task on sooner, rather than later as a decent classification system will allow you to strategically focus on your most valuable systems.

If you find this a daunting task, getting started, I recommend you pick up a book, The Pragmatic CSO by Mike Rothman. An additional resource is ISO IEC 17799 (2000) TRANSLATED INTO PLAIN ENGLISH which gives you a basic plan for classification based on the ISO17799.

Assuming you now know where your critical assets live, your next step is to develop a sound vulnerability management program. What exactly is vulnerability management though? If you do a Google Search it will return hundreds of links with vendor solutions for you. A good starter document is NIST Publication 800-40-Ver2. But in this writer's opinion, patch management is a process that allows us be compliant with policy or regulations, but does not really achieve our goal of securing our systems. In today's world of multiple Zero Day Attacks, patching will not keep us safe.

The strategist needs to know where he is most vulnerable to threats. A vulnerability scanning program is a good place to start, but in environments where staff is already overworked, vulnerability scan reports can go unheeded, or remediation plans un-executed. A seasoned cyber strategist will put in place a pro-active plan for finding vulnerabilities before the bad guys do. This involves scanning web applications and performing penetration testing. The value of discovering vulnerable systems before they are breached can be calculated and used as evidence to management that your program works.

Let's say that through your pro-active penetration testing discover an application that is vulnerable to a SQL Injection which exposes 50,000 records of customer data. Through your investigation, you also show that the first time the data was breached was when your team performed their penetration tests. Using a Ponemon Institute study that places the cost per record to recover from a data breach at $202 per exposed record. You can now demonstrate a cost avoidance of $10.1 million. Metrics like this are valuable tools for the cyber strategist and can be used to justify existing programs or add new ones.

The one challenge to this is that penetration testing is expensive if you hire an outside firm. It is also difficult to find qualified staff in house who have the depth of experience and knowledge to perform successful tests, as well as the bandwidth to perform the tests. There are automated tools available that can mitigate this challenge. One such tool that the cyber strategist uses is Core Impact Pro.

More on this in Part 3, "Your tools."

The Key Of Knowledge (Earth Book) Part 3

Given that the state of cyber security is constantly in flux, our ability to understand our environment and the threats against it hinges on the choice of tools we use. Those tools can be as varied as the number of hackers attempting to compromise our assets, so the strategic use of tools will be my focus, with a few examples of the tools I have found to be the most valuable in multiple dimensions.

Just like in a good cyber defense, a layered approach to situational awareness tools as well. Consider all the sources of information about our environment that are available to us, things such as firewall logs, system security logs, application logs, NetFlow, anti-virus logs, patch statistics, and on and on. Use of an automated tool such as a SIEM is critical for the strategist. The automatic collection of disparate data, collation of events and display of trends, alerts and other information is invaluable to seeing the larger security picture of your environment. Understanding what is going on around you, seeing trends, knowing what is happening in your own environment. In my current environment, our SIEM collects over a billion events a month. It provides real time displays of target information, such as SQL Inject attack sources, bot-net activity, anti-virus alter trends and firewall connections. Charting deviation to normal behavior can alert us real-time to a developing attack or incident. Regrettably, due to funding and staff, the volume of information we collect is not being utilized to its fullest extent, but strategically, the real time views are invaluable.

Another tool that is invaluable strategically is automated penetration testing. I have seen and heard various discussions on the value of penetration testing in the enterprise, but my experience has proven it to be extremely valuable on a number of fronts. The first is as a security awareness tool. Performing vulnerability scans on systems, then using the results to effect change can be problematic at best. The number of false positives, the volume of irrelevant vulnerabilities detected, and the nature of the increasing rapidity in which vulnerabilities are announced can cause your system administrators to put any remediation plans (if they even have the time to make one) on their operational back burner. After all, how many times have you given them similar reports and even though remediation was never complete, their systems continued on without a compromise? Although we know it is just a matter of time until it happens, operational issues trump their view of the risk.

Replay the scene with the use of an automated penetration testing tool. Instead of giving them reports that are meaningless to them, use the results of your vulnerability scans to perform the penetration tests. Identify the real vulnerabilities that matter, and demonstrate how easily their systems are compromised will demonstrate how serious the risk can be. Your strategic knowledge of real risk will educate the admins and bring them on board as partners in securing your enterprise.

The cyber strategist can also use the same tool test the enterprises critical assets. Application attacks have been reported to be on the rise since as long ago as 2004 and earlier. Is the application development life-cycle in your enterprise done with security in mind from the very earliest stages? If, like most enterprises, it comes later in the process, your application infrastructure can pose real risk your enterprise. Imagine the strategic value of discovering a vulnerability in your application that exposes customer or confidential data, and being able to show that you were the first to discover it. The value of the cost-avoidance can be used as a marketing tool to justify or expand your security operation.

The Key Of Knowledge (Earth Book) Part 4

Tools to help you understand, to know your environment are a foundation of this Key, but not the only component. The strategic use of knowledge as a tool to combat cyber attacks is critical to your successful defense. Over the years the attack landscape and corresponding security controls have grown in intensity and complexity. It wasn't so long ago that viruses and web defacement were primary concerns, and firewalls and anti-virus software were the leading controls of self-defense. Those days are gone though, with the ever changing nature and craftiness of attacks.

Zero Day Attacks used to be discussed as something coming in the future, but now it seems they come as fast and furious as rain drops in a tropical storm. When the SQL Slammer hit in January 2003, a patch had already been released 6 months earlier. Now, According to IBM's 2008 X-Force midyear report, more than 90 percent of browser-related exploits detected during the first six months of the year have occurred within 24 hours after these vulnerabilities were disclosed.

The nature of threats continues to develop at an amazing pace. The 1Q 2009 IBM Internet Security Systems X-Force Threat Insight Quarterly reports that the increase of Insider Threats, Web Exploits and exploits such as the PDF exploit that can be triggered simply by displaying the icon in a browser or other application.

Industry certifications are also a component of this key. While there may be some disagreement as the value of certifications, or which certifications to hold, from a strategic viewpoint I believe they are invaluable. And there are a number of reasons why I believe this to be true.

The first reason is the intrinsic value that achieving certification brings to the strategist's psyche, the accomplishment of being acknowledged as a knowledgeable person in your field. The investment that you must make in time studying, the dollars for study materials and the cost of the test itself demonstrate your commitment to self-improvement. Plus the fact that many companies simply add certain certifications as requirements to job descriptions, make the value of certification tangible in a financial way to the strategist.

But how can certifications be used strategically in your day-to-day security responsibilities? Should you choose vendor neutral or technology specific certifications? There have been many articles written on those questions, and you may use this as your first exercise in strategic thinking. The first thing you should identify is what skill set is required to perform the function of your current position. If you are currently a security manager, responsible for an organizational program, perhaps the Certified Information Security Manager (CISM) from ISACA or the Certified Information Systems Security Professional (CISSP) from (ISC)²® would be appropriate, given you have the required experience to qualify for application.

Certifications can also be used as part of your marketing strategy. I have seen on many occasions where part of management culture values educational accomplishments over real world success and expertise, and if you are in such an environment, certifications can go a long way in establishing credibility. Remember, certifications can mean many things to many people, but to the cyber-strategist, they are simply another took in their arsenal of the Key of Knowledge.

The Key Of Knowledge (Earth Book) Part 5

Sharing of knowledge is part of this Key as well. In the industry this is commonly called Security Awareness, but it has been this strategists experience that many of our users are anything but aware. Conventional wisdom dictates that we prepare security awareness training and require our users to complete the course, we hang posters, write and send newsletters, promote security awareness campaigns and hope that the message gets through. And yes, I believe that these are all components to a successful strategy, but are not sufficient in and of themselves. In my environment we do all of those things, with regular updates to keep it fresh and as interesting as possible, but how effective is it really?

In 2009, we saw a significant increase in targeted spear-phishing emails in our enterprise. The goal of the attack was a social engineering attempt to get the user to voluntarily give up their login and password for their email. The number of users who actually fell prey to the attack was extremely small when compared to the enterprise, but the success of the attackers caused the domain to be blacklisted by a number of organizations. The users who gladly sent their passwords and account name off in order to get more in-box storage had all taken and completed the mandatory training, which had a specific module about this very type of attack. Perhaps without the existing training, the number of successfully compromised accounts would have been greater, but that is difficult to quantify. The resulting remediation required a significant amount of effort, but could it have been avoided?

Michael Santarcangelo talks about the difficulty in measuring success of security awareness in his book Into the Breach. He discusses the concept that if there is no adverse effect from and end user violating security practices, then there is no motivation to comply. Perhaps in your organization you have the ability to supply motivation through disciplinary methods, but in the environment where I work, unless there has been some sort of public breach or law violated, corporate discipline is not an available tactic that is available.

So, this begs the question, how does a Security Strategist effectively impart security awareness to the user community? First, don't stop your conventional programs. Although the success metrics may be difficult to measure, I think the programs succeed with a large majority of users. I also believe there are better ways to engage the users in the educational process, methods that clearly demonstrate an adverse effect to the user, that can elicit an emotional response.

There are several methods I have used to engage various communities. The first is the use of a cyber security tabletop exercise, or TTX. Typically these involve security and IT staff, but can include other areas such as management and public relations people. During an exercise participants can experience the results of significant cyber events without affecting the day to day operations of the enterprise. As a veteran player and planner of TTX's, I can tell you that during an exercise, stress and emotions can run high and leave an impression on players, an imprint that will last into their daily operations.

Another method I have found useful is to simulate an event for end users, one that if they fall victim to, will educate them to the potential damage that could have been done. Imagine an end user finding a USB thumb drive, and instead of turning it in, decides to find out what interesting things may be on the drive. When they plug it into their computer, violating policy, they are greeted with a message educating them on the dangers of the action they just performed. The emotion of being caught doing something wrong will insure that user thinks twice before they do something like that again.

There are other methods that may or may not work in your environment, but, the strategist will use the knowledge of their landscape, of their users, their policies, their threats and vulnerabilities and develop security awareness strategies that will impact the users in a positive way, adding another layer of defense in our never ending struggle with those who would compromise our enterprises.

The final part of the Key of Knowledge is Security Awareness. The threat landscape is continually evolving as is the industries response to those threats. According to McAfee Labs 2010 Threat Report "McAfee Labs foresees an increase in threats related to social networking sites, banking security, and botnets, as well as attacks targeting users, businesses, and applications". What I find of particular note is the report is their predictions for 2010 are historical fact in my environment for 2009. Attacks targeting users were the number one source of successful attacks.

The Key of Agility (Water Book)

In the Book of Water Musashi says that water adopts the shape of its receptacle, it is sometimes a trickle and sometimes a wild sea. The book continues with instruction on the use of the sword, spirituality, temperament, attitudes of swordsmanship and balance. When absorbing this book, not merely memorizing it as Musashi says, you lean that balance in all things, and calmness of spirit, and the ability to be agile are critical to the Way. His descriptions of the various methods for the strategic use of the sword in battle. In the world of the cyber strategist, this relates to how we do defend against those attacking our systems, or during our incident response. In his next book, Fire, Musashi takes these principles and applies to the proactive fight, or battle, but like Musashi, the Key Of Agility will focus on defending ourselves.

Musashi talks of the Spiritual Bearing in Strategy, the Stance in Strategy and The Gaze in Strategy. These relate to the cyber strategist in how you have prepared yourself and your enterprise to respond in the time of an attack. When I first assumed the mantle of cyber security leadership in my organization, I analyzed the state of incident response. What I discovered was incident response was better described as incident reaction. When something happened, the reactions were to ask two questions first. Does anyone else know about it? And, do we have to tell anyone else? Then typically, systems were restored and it was business as usual, with no lessons learned, and on occasion not knowing or understanding that the restoration of business did nothing to remove the attacker or threat. This is an example of the lack of a Spiritual Bearing in Strategy. The strategist must develop a process that is based on sound principles.

Holding the Long Sword

End of The Book of 5 Keys

Work halted end of 2010

Appendix G: Actionable Intelligence in Cybersecurity Continuous Monitoring

Bridging the Gap Between Vendor Claims and Operational Reality

1. Introduction: The Ambiguity of "Actionable Intelligence" in Cybersecurity

The term "actionable intelligence" permeates contemporary cybersecurity discourse, serving as a cornerstone concept, particularly for vendors offering threat intelligence platforms and continuous monitoring solutions.[1] Its appeal lies in the promise to elevate an organization's security posture from a reactive state, merely responding to incidents after they occur, to a proactive one, capable of anticipating and mitigating threats before they cause harm.[1] This proactive stance is deemed essential in the face of an increasingly complex and dynamic threat landscape.[20]

However, beneath the surface of this ubiquitous term lies a significant tension, often stemming from differing perspectives on what constitutes "actionable." A rigorous definition, frequently influenced by military and intelligence community doctrine, emphasizes the necessity of human analysis, deep contextualization, and the formulation of tailored recommendations before intelligence can truly be considered actionable and support effective decision-making.[22] Conversely, cybersecurity vendors, particularly those marketing automated solutions, often imply that their platforms deliver intelligence that is immediately "actionable" upon generation, suggesting that the necessary analytical work has already been completed by the system with minimal need for human intervention.[5] This discrepancy centers on the interpretation of "actionable" and the perceived completeness and reliability of the intelligence product delivered by automated systems.

This report aims to dissect this ambiguity and provide clarity on the nuances of "actionable intelligence" within the specific context of cybersecurity continuous monitoring. Its objectives are:

- To define "actionable intelligence" by examining perspectives from the cybersecurity industry (including standards bodies like NIST and analysts like Gartner) and the military/intelligence community.

- To analyze how cybersecurity vendors, especially those offering continuous monitoring solutions, operationalize and market the concept of "actionable intelligence."

- To detail the process involved in transforming raw security data into intelligence suitable for action, with a particular focus on the indispensable role of human analysis and contextualization.

- To compare and contrast the vendor usage of the term with the more stringent military/intelligence definition, highlighting critical differences, especially concerning the human element.

- To explore critiques and discussions within the cybersecurity community regarding the potential overuse or misrepresentation of "actionable intelligence" by vendors.

- To evaluate concrete examples of outputs from vendor tools that are presented as "actionable intelligence."

- To synthesize these findings, clarifying the term's multifaceted nature and addressing the core concern regarding the potential gap between vendor marketing claims and the practical requirements for truly informed action.

The discussion is framed within the context of continuous monitoring, a practice vital for managing security in dynamic IT environments.[5] Continuous monitoring systems generate vast quantities of data from logs, network traffic, endpoints, and other sources.[1] The promise of automatically transforming this data deluge into timely, actionable insights is therefore particularly compelling for organizations seeking real-time risk management.[8] However, this reliance on automation also makes a clear understanding of what constitutes genuinely "actionable" intelligence crucial, as information overload remains a significant challenge.[15] Misinterpreting the nature of automated outputs can lead to ineffective security decisions and a false sense of security.

2.   Foundational   Definitions: What   is   "Actionable" Intelligence"?

Understanding the term "actionable intelligence" requires examining its definition from the perspectives of both the cybersecurity industry, where it is often linked to threat data and vendor solutions, and the military/intelligence community, where it has a longer history tied to operational decision-making.

2.1 The Cybersecurity Perspective: Enabling Informed Decisions

Within cybersecurity, the concept of actionable intelligence is intrinsically linked to Cyber Threat Intelligence (CTI). CTI is broadly understood as the collection, processing, and analysis of data concerning threat actors, their motives, targets, and attack methodologies.[1] The fundamental purpose of CTI is to transform raw data, often voluminous and noisy, into actionable insights that enable security teams and organizational leaders to make informed, data-driven decisions regarding security posture, risk mitigation, and incident response.[1] This transformation aims to shift organizations from merely reacting to attacks to proactively anticipating and defending against them.[1]

Industry analysts like Gartner define threat intelligence as "evidence-based knowledge... [that] provides context, mechanisms, indicators, and action-oriented advice on both existing and emerging threats".[1] This definition underscores that actionable intelligence is more than just data points (indicators); it incorporates context and, crucially, "action-oriented advice," implying a level of interpretation and guidance necessary to direct a response.

Standards bodies like the National Institute of Standards and Technology (NIST), particularly in the context of Security Information and Event

Management (SIEM) systems, define a SIEM tool as an application that gathers security data and presents it as "actionable information via a single interface".[28] While highlighting the importance of usability and centralized presentation, this definition is less explicit about the depth of analysis or contextualization required for information to be truly considered "actionable intelligence" capable of driving effective decisions. It focuses more on the output format than the analytical rigor behind it.

Cybersecurity vendors and practitioners frequently define actionable threat intelligence as distilled, contextual, and timely data that empowers security teams to identify, prioritize, and mitigate risks effectively.[2] Key attributes often cited include being specific to the organization's unique environment (attack surface, vulnerabilities, assets), detailed (covering threat actors, TTPs, IoCs), contextual, and directly enabling action.[2] Relevance, timeliness, and accuracy are paramount.[3] A primary function is to cut through the "noise" of excessive alerts and raw data, allowing teams to focus on genuine threats.[9] This intelligence is intended to support a wide range of security functions, including proactive threat detection, incident response, vulnerability management, risk analysis, and strategic security investments.[1]

Despite variations in emphasis, a common thread emerges across these cybersecurity definitions: actionable intelligence is fundamentally about supporting *better decision-making*.[1] Whether for a SOC analyst responding to an alert, a vulnerability management team prioritizing patches, or a CISO allocating resources, the intelligence provided should lead to more effective choices. The core ambiguity, however, lies not in this purpose but in the *process* required to achieve it. Specifically, how much analysis, interpretation, and human judgment are necessary before data or information can confidently support a security decision? Vendor marketing often emphasizes the role of automation in delivering these insights, while other definitions and practitioner discussions highlight the critical role of human analysis, particularly for more complex intelligence types.[1] This raises the question of whether the output of many automated systems is truly "actionable intelligence" in the sense of a fully formed recommendation, or rather "actionable information" – a valuable input that still requires significant human cognitive processing to inform a final decision.

2.2 The Military/Intelligence Community Perspective: Enabling Successful Operations

The military and intelligence communities have a long-established concept of actionable intelligence, shaped by the demands of operational planning and execution in complex and often hostile environments. Here, actionable intelligence is defined as providing commanders and soldiers with a "high level of shared situational understanding," delivered with the necessary "speed, accuracy, and timeliness" to enable the planning and conduct of "successful operations."[22] The core purpose is to enable commanders to effectively employ their available resources (combat power) to achieve mission objectives, including identifying and targeting adversary vulnerabilities.[23]

Several criteria are consistently emphasized as essential for intelligence to be considered actionable in this context:

- **Timeliness:** Information must be current to be relevant in fast-moving operational situations.[22] Old intelligence is rarely useful.[37]

- **Accuracy:** Decisions must be based on reliable, validated information.[22]

- **Relevance:** Intelligence must directly address the commander's critical information needs (often formalized as Priority Intelligence Requirements or PIRs) and be pertinent to the specific operational context.[24]

- **Usability & Completeness:** Intelligence must be presented in a format that the recipient can understand and use, providing sufficient detail to support decision-making.[38]

- **Objectivity:** Intelligence analysis should strive for unbiased assessment.[38]

- **Discoverability:** Relevant intelligence should be accessible to those who need it.[38]

A key distinction in the military perspective is the tight coupling between intelligence and the decision-maker's needs and the ultimate operational outcome.[22] Intelligence is not generated for its own sake but is specifically tailored to answer critical questions (PIRs) that inform the commander's plan and actions.[37] It must provide *situational understanding*, which extends beyond simply identifying a threat to encompass a comprehensive grasp of the operational environment, including adversary capabilities, intentions, disposition, and potentially broader factors like terrain, weather, and socio-cultural dynamics.[22]

While technology plays a crucial role in collection, processing, and dissemination (e.g., through systems like DCGS-A or integrated sensor networks),[22] the human element remains central. Human intelligence (HUMINT) collection is often vital, particularly in asymmetric or counter-insurgency warfare where understanding human networks and intentions is paramount.[22] The concept of "Every Soldier is a Sensor" (ES2) further highlights the importance of distributed human observation and reporting.[22] Furthermore, human analysts are critical for interpreting collected information, assessing adversary intent, integrating data from multiple sources (all-source intelligence), and providing the nuanced understanding required by commanders.[23]

From this perspective, intelligence becomes "actionable" when it directly contributes to the commander's ability to make a sound operational decision that increases the likelihood of mission success. It requires a level of analysis and contextualization that provides not just data, but understanding relevant to a specific operational problem. The threshold for actionability is therefore determined by its *operational utility* – its ability to inform the best course of action within a specific context – rather than merely its technical usability or potential to trigger an automated response. This focus on deep situational understanding and direct support to operational outcomes sets a potentially higher bar for "actionability" than often implied in cybersecurity marketing.

3. From Data to Decision: The Intelligence Transformation Process

The transformation of raw data into actionable intelligence is not an instantaneous event but a structured, cyclical process. Understanding this process, particularly the critical analysis stage and the role of human expertise, is essential for appreciating the nuances of "actionable intelligence."

3.1 The Threat Intelligence Lifecycle in Cybersecurity

The cybersecurity industry generally adopts a model known as the threat intelligence lifecycle, a continuous and iterative process designed to systematically collect, process, analyze, and disseminate intelligence.[1] While variations exist, the core stages typically include:[1]

1. **Requirements (or Planning & Direction):** This foundational stage involves defining the goals and objectives of the intelligence program. It requires understanding stakeholder needs (from SOC analysts to executives), identifying critical assets and potential attack surfaces, researching potential adversaries and their motivations, and formulating specific questions that intelligence gathering should answer.[1] These requirements guide the entire process and ensure that intelligence efforts are aligned with organizational priorities and risk management objectives.[27] This stage mirrors the military concept of defining Priority Intelligence Requirements (PIRs) to focus collection efforts.[37]

2. **Collection:** Based on the defined requirements, raw data is gathered from a wide array of sources. These can include internal sources like network traffic logs, SIEM alerts, EDR telemetry, vulnerability scan results, and incident reports, as well as external sources such as public threat feeds, open-source intelligence (OSINT) from news articles, social media, forums, and blogs, dark web monitoring, commercial intelligence services, and information shared through communities like ISACs.[1] Human sources may also contribute.[1]

3. **Processing:** Raw collected data is often unstructured, redundant, or in disparate formats. This stage involves transforming it into a usable format suitable for analysis. Activities include organizing information, decrypting files, translating languages, standardizing data formats (e.g., into spreadsheets or databases), filtering out irrelevant or duplicate data, and evaluating the reliability and credibility of the sources and information.[1] Automation, including AI and machine learning techniques, is frequently employed in this stage to handle large volumes of data efficiently.[6]

4. **Analysis:** This is the core stage where processed information is converted into intelligence. Analysts examine the data to identify patterns, trends, and anomalies; correlate information from different sources; assess the capabilities, intentions, and TTPs of threat actors; determine the potential impact of threats on the organization; and ultimately generate actionable insights and recommendations to answer the questions posed in the requirements phase.[1] Contextualization is a key activity within analysis.

5. **Dissemination:** The analyzed intelligence is delivered to the relevant stakeholders in a clear, concise, and understandable format tailored to their specific needs and technical expertise.[1] Outputs can range from detailed reports for executives, technical briefings for security teams, real-time alerts pushed to SOC consoles, or intelligence feeds integrated directly into security tools like SIEMs, SOAR platforms, firewalls, or EDR systems.[6]

6. **Feedback:** The lifecycle is closed by gathering feedback from the consumers of the intelligence. This input helps evaluate the effectiveness, relevance, and timeliness of the intelligence provided and is used to refine future requirements, collection strategies, analysis techniques, and dissemination methods, ensuring continuous improvement of the process.[1]

3.2 The Crucial Analysis Stage: Adding Context and Meaning

The analysis stage is where mere data points are imbued with meaning and transformed into intelligence. It moves beyond simple indicators of compromise (IoCs) – such as IP addresses, domain names, file hashes, or registry keys – to provide a deeper understanding of the threat landscape.[2] Effective analysis seeks to answer the fundamental questions of *who* is attacking, *what* are their capabilities and targets, *why* are they attacking (motivation), *when* and *where* might they strike, and *how* do they operate (TTPs).[1] This involves correlating disparate pieces of information – for example, linking a suspicious IP address (IoC) to a known threat actor group, their documented TTPs, and a currently active campaign targeting the organization's industry.[2]

Context is the critical ingredient added during analysis.[1] Raw data, such as an alert about a vulnerability, lacks context on its own. Analysis provides this context by considering factors such as:

- **Threat Actor Context:** Understanding the adversary's TTPs, motivations (e.g., financial gain, espionage, disruption), capabilities, infrastructure, and historical targets.[1]

- **Organizational Context:** Relating the threat information to the specific organization's environment – its industry sector, geographic location, critical assets, known vulnerabilities, existing security controls, and overall risk posture.[2] Analysis of internal data alongside external threat feeds is crucial for creating this "contextual CTI".[4]

- **Impact Context:** Assessing the potential business impact if the threat materializes, considering factors like financial loss, operational disruption, reputational damage, regulatory penalties, and data breach consequences.[4]

- **Temporal and Landscape Context:** Understanding how a specific threat fits into broader trends, current events, geopolitical situations, and the overall evolving threat landscape.[1]

A key outcome of applying context during analysis is effective prioritization.[3] Security teams face a constant stream of alerts and potential threats but have limited resources.[19] Analysis helps prioritize which threats pose the greatest risk based on factors like likelihood of exploitation, potential impact, asset criticality, and relevance to the organization, allowing teams to focus their efforts where they matter most.[12]

Ultimately, analysis involves interpretation, deduction, and judgment.[1] It transforms processed information into actionable insights – conclusions drawn from the evidence – and often includes specific recommendations for mitigation, remediation, or strategic adjustments.[1]

3.3 The Irreplaceable Role of Human Expertise

While automation, artificial intelligence (AI), and machine learning (ML) are increasingly employed in the intelligence lifecycle, particularly for processing vast datasets, correlating known patterns, and automating routine tasks, human expertise remains indispensable, especially in the critical analysis stage.[6]

Automation excels at speed and scale, efficiently handling known threats and patterns identified through signatures, rules, or algorithms. However, it often falls short when dealing with ambiguity, novelty, and the need for deep contextual understanding. Human analysts are crucial for several reasons:

- **Deeper Analysis and Interpretation:** For operational intelligence (understanding the 'who, why, how' of specific attacks) and strategic intelligence (high-level threat landscape and risk analysis), human analysis is often explicitly required to convert data into meaningful insights.[1] These levels of intelligence typically require interpreting complex situations, understanding adversary motivations, and assessing strategic implications – tasks that demand human cognitive abilities.[1]

- **Handling Novelty and Ambiguity:** Humans possess the critical thinking skills needed to analyze novel threats, such as zero-day exploits or new TTPs for which no automated signatures exist. They can work with incomplete or ambiguous information, make inferences, and apply intuition and experience to assess situations that fall outside predefined patterns.[1] Understanding adversary intent, which often requires interpreting subtle clues or understanding cultural or geopolitical context, is another area where human judgment is vital.[1] Strategic intelligence, in particular, often necessitates human expertise in both cybersecurity and relevant domains like geopolitics.[1]

- **Validation, Curation, and Quality Control:** Human analysts play a vital role in validating the outputs of automated systems, filtering out false positives that can overwhelm security teams, and assessing the credibility and reliability of information sources.[34] They curate intelligence, ensuring its relevance and tailoring it to the specific needs and understanding of different audiences within the organization.[36]

- **Tailoring Recommendations:** Developing nuanced, actionable recommendations that consider the organization's unique technical environment, business context, risk tolerance, resource constraints, and strategic goals typically requires human judgment and expertise.[1] Generic recommendations may not be optimal or even feasible for a specific organization.

The military's continued reliance on HUMINT analysts and all-source intelligence professionals underscores the enduring value of human cognition in understanding complex, dynamic, and adversary-driven environments.[22] While technology provides powerful tools for data collection and processing, it is the human analyst who often provides the critical layer of interpretation, contextualization, and judgment needed to transform that data into reliable, actionable intelligence suitable for high-stakes decision-making. Automation is adept at processing knowns and matching patterns, but human expertise acts as the essential "context engine," interpreting the significance of information, navigating uncertainty, and tailoring insights for specific decisions and strategic actions, especially those extending beyond simple, immediate technical responses.

4. Vendor Landscape: "Actionable Intelligence" in Continuous Monitoring Tools

The concept of "actionable intelligence" is central to the marketing and value proposition of many cybersecurity vendors, particularly those offering continuous monitoring solutions, threat intelligence platforms (TIPs), SIEMs, and related services. Understanding how these vendors position and deliver on this promise is crucial for practitioners seeking to leverage these tools effectively.

4.1 Marketing vs. Reality: How Vendors Position "Actionable Intelligence"

Cybersecurity vendors consistently emphasize their ability to provide "real-time, actionable intelligence" as a key differentiator.[5] Marketing materials frequently highlight attributes such as:

- **Speed and Real-Time Delivery:** Promising immediate insights into emerging threats, vulnerabilities, or compromises.[5]

- **Automation:** Touting the use of AI, machine learning, and automated processes to collect, process, analyze, and deliver intelligence, often implying reduced manual effort for security teams.[6]

- **Noise Reduction:** Claiming to filter out irrelevant data and false positives, delivering only high-fidelity, relevant alerts that require attention.[5]

- **Proactive Defense:** Positioning their solutions as enabling organizations to move beyond reaction and proactively detect, prioritize, and mitigate risks before incidents occur.[1]

- **Integration:** Highlighting seamless integration with existing security infrastructure, such as SIEM, SOAR, EDR, and firewalls, to enable automated responses or enrich investigations.[6]

Within the domain of continuous monitoring, "actionable intelligence" is presented as the crucial output that makes sense of the constant stream of data generated by monitoring tools. Continuous Controls Monitoring (CCM) platforms, for example, aim to provide 24/7 visibility into control effectiveness, risk posture, and compliance status, often promising a "single source of truth" derived from analyzing data across disparate tools.[25] These platforms leverage technologies like AI and machine learning to detect anomalies, deviations, coverage gaps, or misconfigurations, presenting these findings as actionable insights for remediation.[8]

A common theme in vendor marketing is the implication that the platform itself performs the necessary analysis and delivers a finished intelligence product that is ready for action. Language such as "providing real-time, actionable intelligence by continuously monitoring credentials... while instantly mapping affected assets,"[5] "delivers real-time, concise, and actionable alerts,"[6] or providing "context-specific actionable intelligence (CSAI) to perform... automated attack surface analysis,"[26] suggests a high degree of automation in the analysis and interpretation phases. While some descriptions acknowledge that flagged items may require "further investigation,"[25] the emphasis is often on the automated delivery of insights deemed "actionable" by the system.

The focus of this marketed "actionable intelligence" is frequently on specific, often technical, outputs. Vendors highlight their ability to detect compromised credentials being sold on the dark web,[5] identify exploitable vulnerabilities on the attack surface,[8] surface specific IoCs associated with active campaigns,[18] detect malware or phishing attempts,[9] or identify misconfigurations.[26] These are presented as concrete findings that enable immediate defensive actions.

4.2 Evaluating Vendor Outputs: What Do Alerts Typically Provide?

When examining the actual outputs generated by continuous monitoring tools, TIPs, SIEMs, and related platforms that are labeled as "actionable intelligence," a spectrum of information types and analytical depth becomes apparent. Common examples include:

- **Indicators of Compromise (IoCs):** These are perhaps the most frequent form of output presented as actionable. They consist of technical artifacts associated with malicious activity, such as IP addresses of command-and-control servers, domains hosting phishing sites, file hashes of known malware, or malicious URLs.[2] These are often delivered via threat feeds designed for automated ingestion by security tools (firewalls, SIEMs, EDR) to block known threats or trigger alerts.[4]

- **Vulnerability Information:** Alerts identifying vulnerabilities (often referenced by CVE numbers) present in an organization's systems or software.[3] More advanced platforms may link these vulnerabilities to specific assets identified through scanning or asset management, indicate if the vulnerability is being actively exploited in the wild, and provide a risk score or prioritization level based on factors like exploitability, potential impact, or asset criticality.[12]

- **Threat Actor Information:** Some platforms provide intelligence on specific threat actors or groups, including their known TTPs, motivations, targeted industries, and associated campaigns or IoCs.[1] This is often delivered through reports or profiles within the platform.

- **Correlated Alerts:** SIEM, XDR (Extended Detection and Response), and TIPs often correlate events from multiple security tools and data sources to identify potential incidents that might be missed when looking at individual alerts in isolation.[9] For example, correlating a firewall alert for outbound traffic to a suspicious IP, a SIEM alert for a login from an unusual location, and an EDR alert for data movement from the same endpoint could generate a high-priority incident alert.[50]

- **Specific Threat Alerts:** Tools specializing in areas like dark web monitoring, data leak detection, or attack surface management generate alerts for specific events like leaked employee credentials found online,[5] sensitive data exposure on misconfigured cloud storage,[35] active phishing campaigns targeting the organization,[9] or detection of botnet activity.[35]

- **Risk Scores and Prioritization:** Many tools assign numerical risk scores to vulnerabilities, assets, or alerts based on various factors, aiming to help security teams prioritize their remediation or investigation efforts.[12]

The level of analysis and context provided with these outputs varies significantly:

- **Low Context:** Raw IoC feeds or basic vulnerability alerts often provide minimal context. A list of malicious IPs, for instance, requires significant human effort to determine its relevance to the organization, assess the potential impact, and decide on the appropriate action beyond simple blocking.[49]

- **Medium Context:** Outputs become more valuable when enriched with additional context. This includes correlated alerts that link events across different systems,[50] alerts enriched with data from threat intelligence feeds (e.g., identifying an IP as belonging to a known ransomware group),[9] or vulnerability alerts prioritized based on exploitability data and internal asset criticality.[26] An alert showing communication between a critical internal server and an IP address known to be malicious represents a higher level of contextualized information.[18]

- **High Context:** The highest level involves detailed analysis and tailored recommendations. This often takes the form of comprehensive reports on threat actors, campaigns, or specific incidents, providing deep insights into TTPs, motivations, potential impact, and specific mitigation strategies relevant to the organization's environment.[1] Generating this level of intelligence typically involves significant human analyst effort, either from the vendor's intelligence team or the organization's internal team interpreting lower-level outputs.

Considering this spectrum, much of what vendors label "actionable intelligence," particularly the automated outputs from continuous monitoring systems, aligns more closely with what could be termed *technically actionable information* or *prioritized alerts*. These outputs provide a necessary starting point for action – they indicate *that* something requires attention (e.g., block this IP, patch this CVE, investigate this user activity). They are "actionable" in a technical or tactical sense. However, they often lack the comprehensive analysis, deep organizational context, impact assessment, and tailored strategic recommendations characteristic of the more rigorous military/intelligence definition of actionable intelligence. The "action" enabled is frequently immediate and tactical, rather than strategic and fully considered based on a complete understanding of the situation. This observation aligns with the user's initial query regarding the discrepancy between vendor claims and a definition requiring deeper human analysis to achieve true decision support.

5. Bridging the Gap: Comparing Vendor Usage and Military/Intelligence Doctrine

The differing interpretations of "actionable intelligence" between cybersecurity vendor marketing, particularly in the continuous monitoring space, and the established doctrine of military/intelligence communities lead to significant contrasts in emphasis and expectation. Understanding these differences is key to navigating the term's practical application.

5.1 Key Differences Highlighted

A direct comparison reveals several fundamental points of divergence:

- **Emphasis on Human Analysis:** This is perhaps the most critical distinction. Military and intelligence doctrine views human analysis, judgment, and interpretation as essential components for developing situational understanding and generating intelligence that can reliably inform command decisions, especially in complex or novel situations.[22] While acknowledging the role of technology, the human analyst is central. Vendor marketing, conversely, often emphasizes the power of automation (AI/ML) to process data and deliver insights, sometimes implying that human analysis is minimized, supplementary, or only required for exceptional cases.[5] While some vendor literature and more nuanced discussions do acknowledge the necessity of human analysts for certain intelligence types (like operational or strategic CTI),[1] the prevailing marketing message often leans heavily on automation delivering the "actionable" output.

- **Definition of "Action":** The military perspective implies that "action" is taken to achieve a specific operational objective or mission success, based on a comprehensive understanding of the situation.[22] The intelligence must inform the *best* course of action. The vendor perspective, particularly regarding automated alerts from continuous monitoring, often implies more immediate, technical actions: blocking an IP address, patching a vulnerability, isolating a compromised machine, or triggering a SOAR playbook.[9] The focus is on enabling a rapid, often tactical, response.

- **Depth of Context:** Military intelligence strives for deep, holistic situational understanding, incorporating adversary capabilities and intent, the broader operational environment, and potentially political or cultural factors.[22] Vendor-provided outputs, especially automated alerts, vary greatly in contextual depth. While some platforms offer enrichment by correlating internal data or external feeds,[4] achieving the level of deep organizational or strategic context often requires significant additional human analysis or investment in high-end platforms or services.[34]

- **Output Focus and Audience:** Military intelligence products are ultimately geared towards informing a commander's decision-making process.[22] Vendor outputs from continuous monitoring tools are frequently designed for consumption by technical security teams (SOC analysts, incident responders, vulnerability managers) or direct integration into automated security systems (SIEM, SOAR, firewalls).[28] The format and content reflect this difference – military intelligence might be a detailed assessment or briefing, while vendor output is often an alert, an IoC list, or a vulnerability score.

- **Role of Requirements:** The military intelligence cycle is strongly driven by the commander's specific, prioritized information needs

(PIRs).[37] While cybersecurity also emphasizes defining requirements,[1] vendor tools might generate alerts based on generic threat signatures, anomaly detection algorithms, or broad threat feeds unless carefully configured, tuned, and potentially augmented with custom rules based on the organization's specific context and priorities.

5.2 Comparative Analysis Table

The following table summarizes the key differences in how "actionable intelligence" is typically framed within the vendor/continuous monitoring context versus the military/intelligence community doctrine:

Attribute	Vendor Perspective (Continuous Monitoring Focus)	Military/Intelligence Perspective
Definition Focus	Enabling rapid detection, prioritization, and technical response to threats, vulnerabilities, or anomalies [5]	Enabling successful operations through comprehensive situational understanding and informed command decisions [22]
Role of Human Analysis	Often minimized in marketing emphasis on automation; seen as necessary for validation, deeper investigation, or higher-level intelligence[1]	Central and essential for interpretation, contextualization, assessing intent, handling ambiguity, and tailoring recommendations [22]

Primary Output	Alerts, IoCs, vulnerability data, risk scores, correlated events, automated reports [28]	Assessments, briefings, tailored reports, answers to PIRs, situational awareness products [22]
Implied Action	Immediate technical/tactical response (block, patch, isolate, investigate alert) [9]	Strategic or operational decisions and actions aimed at achieving mission success [22]
Key Criteria	Timeliness, relevance (often technical), accuracy, automation, integration [3]	Timeliness, accuracy, relevance (to PIRs/mission), completeness, usability, objectivity [22]
Context Level	Varies; often technically focused (IoC context, vulnerability exploitability); deeper organizational/strate gic context may require more effort [34]	Deep situational understanding, including adversary intent, capabilities, and broader operational environment [22]

5.3 Implications of the Discrepancy for Security Teams

This difference in perspective and definition is not merely academic; it has practical implications for cybersecurity teams utilizing vendor tools:

- **Risk of Misunderstanding and Misaligned Expectations:**
  Security teams might procure and implement tools expecting fully
  analyzed, context-rich intelligence ready for strategic use, based on
  marketing claims. They may then find the outputs are primarily
  technical alerts or prioritized data points requiring significant
  internal effort to interpret and act upon strategically.[54] This can lead
  to frustration and perceived underperformance of the tool.

- **Potential for Suboptimal Actions:** Acting solely on technically
  focused alerts without sufficient contextual analysis can lead to
  inefficient or even counterproductive responses. For example,
  repeatedly blocking IP addresses associated with a content delivery
  network (CDN) flagged in a generic feed might disrupt legitimate
  traffic. Chasing individual alerts might obscure the view of a larger,
  coordinated attack campaign where a more strategic response is
  needed.

- **Resource Drain and Alert Fatigue:** If vendor tools generate a high
  volume of alerts labeled "actionable" but lacking deep context or
  prioritization relevant to the specific organization, security analysts
  can become overwhelmed.[19] Investigating numerous low-fidelity
  alerts consumes valuable analyst time and contributes to alert
  fatigue, potentially causing critical alerts to be missed.

- **Necessity of Internal Processes and Expertise:** The gap between
  vendor-provided "actionable information" and truly "actionable
  intelligence" highlights the critical need for organizations to invest
  in their own internal analysis capabilities. This includes establishing
  clear processes for alert triage, enrichment, investigation, and
  decision-making, as well as ensuring they have personnel with the
  necessary skills and time to perform these tasks.[34] Relying solely on
  the tool's output without this internal capacity is insufficient for
  robust security.

6. Community Perspectives: Critiques of the "Actionable Intelligence" Label

Within the cybersecurity community, the widespread use of the term "actionable intelligence" by vendors has not gone unnoticed or uncriticized. Discussions often revolve around the term's potential for overuse, misrepresentation, and the practical challenges it creates for security practitioners.

6.1 The "Buzzword" Problem: Overuse and Dilution

Similar to terms like "APT" (Advanced Persistent Threat) or "AI-powered," "actionable intelligence" risks becoming a diluted marketing buzzword.[36] When nearly every vendor claims to provide it, the term loses its specific meaning and impact. Vendors may leverage the term primarily to make their products or services appear more sophisticated and valuable, potentially exaggerating the level of analysis or immediate usability of their outputs.[54]

This overuse creates confusion in the marketplace, making it challenging for security professionals to accurately assess and compare different offerings.[36] A simple label of "actionable intelligence" provides little insight into the actual nature of the output. Critical questions remain: Actionable *for whom*? Actionable *for what specific purpose*? Actionable *based on what level of analysis and context*?.[36] Without clear answers to these questions, the term offers limited practical value in evaluating a solution's true capabilities.

6.2 Risks of Misrepresentation and Over-Reliance

The potentially misleading nature of the term carries tangible risks for organizations:

- **Resource Misallocation:** If security teams treat all vendor outputs labeled "actionable" as equally important or fully vetted, they risk misallocating critical resources.[54] Time might be spent chasing down low-impact alerts or patching vulnerabilities that are theoretically exploitable but not actively targeted against the organization, while more pressing, context-specific threats are overlooked. Prioritization based solely on vendor-supplied scores or labels, without internal validation and contextualization, can be flawed.[54]

- **Alert Fatigue and Noise:** The promise of actionable intelligence can sometimes translate into a high volume of alerts that, while perhaps technically accurate, lack sufficient context or relevance to be truly actionable without significant further investigation.[19] Vendors may deliver vast feeds of IoCs or vulnerability data, but without effective filtering, prioritization, and contextualization tailored to the client's specific environment, this can simply add to the noise that security teams struggle with.[53] Constant notifications about risks that do not directly impact the recipient or require immediate action can lead to desensitization, causing genuine critical alerts to be ignored.[52]

- **False Sense of Security:** An over-reliance on automated tools marketed as providing "actionable intelligence" can foster a dangerous sense of complacency.[54] Organizations might believe they are adequately protected simply because they have deployed such tools, underestimating the need for ongoing human oversight, critical thinking, and validation.[52] Believing a tool has successfully defended against a threat mislabeled as highly sophisticated (e.g., an APT) might lead to an inaccurate assessment of preparedness.[54]

- **Distraction from Foundational Security:** An excessive focus on acquiring the latest "actionable intelligence" feeds or platforms might distract organizations from implementing and maintaining fundamental security hygiene practices. Research indicates that attackers frequently exploit older, known vulnerabilities that remain unpatched, sometimes for years.[53] Ensuring robust basics like patch management, secure configuration, and access control remains critical, regardless of the sophistication of intelligence inputs.

6.3 The Need for Critical Evaluation

Given these critiques and risks, the cybersecurity community emphasizes the need for practitioners to adopt a critical stance when evaluating vendor claims about "actionable intelligence."[36] Instead of accepting the label at face value, security teams should probe deeper:

- What specific analysis does the tool or service perform?

- What level of context (technical, organizational, strategic) is provided with the outputs?

- What specific action(s) does the output directly enable?

- Is human validation, interpretation, or further analysis required before a confident decision can be made?[36]

- How is the intelligence tailored or prioritized for *our* specific organization, industry, and risk profile?[2]

It is essential to understand the limitations of different types of intelligence feeds and tools.[53] A generic IoC feed might be useful for automated blocking but provides little strategic insight. A platform excelling at vulnerability prioritization might not offer deep threat actor analysis. Organizations must assess their specific needs and determine what kind of intelligence will provide the most value, recognizing that not every organization requires the same level or type of external intelligence feed.[53]

Ultimately, while vendors market "actionable intelligence" as a solution that simplifies the security team's burden, the reality is often more complex. The ambiguity surrounding the term, combined with the frequent lack of deep, tailored context in automated outputs, can inadvertently shift the burden of performing the *actual* intelligence analysis – the contextualization, interpretation, and validation required for strategic decision-making – back onto the consuming organization's security team. The promise of simplification may mask the reality of a continued, significant need for internal analytical effort to transform vendor-provided information into intelligence that is truly actionable in a comprehensive, strategic sense.

7. Synthesis and Conclusion: Navigating "Actionable Intelligence" in Practice

The term "actionable intelligence" in cybersecurity continuous monitoring represents a complex concept with interpretations that vary significantly between vendor marketing and the more rigorous definitions rooted in military and intelligence community practices. Navigating this landscape requires a nuanced understanding of what is being offered versus what is truly needed for effective security decision-making.

7.1 Understanding the Spectrum: From Raw Alerts to True Intelligence

It is clear that "actionable intelligence" exists on a spectrum rather than representing a single, monolithic concept.

- At one end of this spectrum lies **technically actionable information**. This includes raw or lightly processed data points such as IoCs (IPs, hashes, domains), basic vulnerability alerts, or correlated event logs generated by continuous monitoring systems, SIEMs, or basic threat feeds.[34] This information is often "actionable" in the sense that it can trigger an automated response (e.g., blocking an IP via firewall integration, isolating a host via SOAR playbook) or prompt an immediate tactical action by a security analyst (e.g., initiating a patch, starting an investigation). Its primary value lies in speed and enabling rapid, often automated, tactical responses.

- At the other end lies **strategically actionable intelligence**. This aligns more closely with the military/intelligence community concept.[22] It represents deeply analyzed, contextualized information that provides situational understanding, assesses potential impact specific to the organization, considers adversary intent and capabilities, and directly supports informed strategic or complex operational decision-making.[1] Generating this level of intelligence typically requires significant human analysis, interpretation, and judgment.

Most outputs from automated continuous monitoring tools and standard threat intelligence feeds fall closer to the "technically actionable information" end of the spectrum. They provide valuable signals and starting points but usually require further human-driven processing, analysis, and contextualization to evolve into true strategic intelligence.

7.2 Addressing the User's Concern: Affirming the Necessity of Human Analysis

The core concern regarding the discrepancy between vendor claims and a more rigorous definition involving human analysis is valid. Achieving high-confidence, strategically actionable intelligence – the kind needed for complex decisions beyond immediate technical blocking or patching – almost invariably necessitates the involvement of skilled human analysts.[1]

Automation, AI, and ML are undeniably powerful *enablers* within the intelligence lifecycle.[6] They excel at processing vast amounts of data at speed, identifying known patterns, correlating events across diverse sources, and automating repetitive tasks. This significantly enhances the efficiency and scope of intelligence operations. However, current technology generally does not replace the need for human cognition in areas requiring:

- **Deep Contextualization:** Relating threat data to the unique business operations, risk appetite, and strategic goals of the organization.

- **Interpretation of Ambiguity and Novelty:** Analyzing incomplete information, understanding adversary intent, assessing new or evolving TTPs, and incorporating geopolitical or cultural nuances.

- **Critical Judgment and Validation:** Evaluating source credibility, filtering false positives, assessing the true significance of correlated events, and validating automated findings.

- **Tailored Recommendation Development:** Formulating specific, practical, and prioritized courses of action suited to the organization's specific circumstances.

Therefore, the perspective rooted in military/intelligence doctrine, which emphasizes the centrality of human analysis for achieving actionable situational understanding, remains highly relevant in the cybersecurity domain, particularly for intelligence intended to drive more than just automated technical responses.

7.3 Recommendations for Practitioners

Organizations seeking to effectively leverage continuous monitoring and threat intelligence should adopt a pragmatic approach:

1. **Critically Evaluate Tools and Services:** Look beyond the "actionable intelligence" label. Assess solutions based on the *specific* outputs they provide, the *level* of analysis performed by the tool versus required by the user, the *depth* of context offered, and the *type* of action directly enabled.[36] Understand the integration capabilities and the requirements for tuning and configuration to maximize relevance.

2. **Integrate Human Expertise into Workflows:** Design security operations and incident response processes that explicitly incorporate human review, analysis, and validation stages for alerts and intelligence received from automated tools.[34] Allocate sufficient time and resources for analysts to perform this critical thinking, rather than solely focusing on clearing alert queues. Foster analytical skills within the team.

3. **Define Internal Intelligence Requirements:** Establish clear, prioritized intelligence requirements (analogous to military PIRs) based on the organization's specific risk landscape, critical assets, regulatory obligations, and strategic objectives.[1] Use these requirements to guide tool selection, configuration, tuning, and the focus of internal analysis efforts.

4. **Prioritize Contextualization:** Invest in tools and processes that

enrich threat data and alerts with relevant organizational context. This includes integrating threat intelligence with asset management, vulnerability data, identity information, and business process understanding to better assess relevance and potential impact.[1]

5. **Actively Manage Alert Volume and Quality:** Implement robust alert tuning, correlation rules, and prioritization mechanisms to combat alert fatigue.[19] Focus analysts' attention on high-fidelity, context-rich alerts that genuinely warrant investigation, rather than drowning them in low-value noise. Continuously refine rules and thresholds based on feedback and operational experience.

7.4 Final Thoughts: The Human-Machine Partnership

Ultimately, achieving genuinely actionable intelligence in the complex and dynamic field of cybersecurity is not a matter of choosing between automation and human expertise, but of forging an effective human-machine partnership.[34] Technology provides the indispensable speed and scale required to collect and process the overwhelming volume of security data and detect known patterns or anomalies. Humans provide the critical thinking, contextual understanding, interpretation of novelty, and strategic judgment necessary to transform that processed data into meaningful insights that drive effective security decisions.

The goal for organizations should be to leverage continuous monitoring tools and threat intelligence platforms not as replacements for human analysts, but as powerful force multipliers that augment their capabilities. By critically evaluating vendor claims, building robust internal processes, and ensuring that technology serves to empower human judgment, organizations can move closer to transforming data overload into the focused, reliable, and truly actionable intelligence needed to navigate the modern threat landscape effectively, aligning practice more closely with the rigorous standard implied by the military and intelligence community definition.

Works cited

Can be found at https://www.c-ooda.com/book-of-five-keys

www.ingramcontent.com/pod-product-compliance
Lightning Source LLC
Chambersburg PA
CBHW051205200326
41519CB00025B/7011